INTERSECTIONS

HISTORIES OF ENVIRONMENT, SCIENCE, AND TECHNOLOGY IN THE ANTHROPOCENE

Sarah Elkind and Finn Arne Jørgensen, *Editors*

UNIVERSITY *of* PITTSBURGH PRESS

FAR BEYOND THE

MOON

FAR BEYOND THE

MOON

A HISTORY OF LIFE SUPPORT SYSTEMS IN THE SPACE AGE

DAVID P. D. MUNNS AND KÄRIN NICKELSEN

Published by the University of Pittsburgh Press, Pittsburgh, Pa., 15260

Manufactured in the United States of America
Printed on acid-free paper
10 9 8 7 6 5 4 3 2 1

Cataloging-in-Publication data is available from the Library of Congress

ISBN 13: 978-0-8229-4654-0
ISBN 10: 0-8229-4654-8

Cover art: Image in window is a drawing of a lunar farm, used under a Creative Commons license from the Space Sciences Institute. Available at http://ssi.org/space-art/ssi-sample-slides/.

Cover design: Alex Wolfe

It was an algae space race.

—William Oswald

Everything is different in a closed system.

—Kim Stanley Robinson, *Red Mars*

CONTENTS

FAR BEYOND THE
MOON

INTRODUCTION

My asshole is doing as much to keep me alive as my brain.

—Astronaut Mark Watney, in *The Martian*

IN 2016 NASA ANNOUNCED THE SO-CALLED SPACE POOP CHALLENGE. It delighted children and amused journalists, who flocked to NASA's Johnson Space Center to hear about the competition. The Poop Challenge called for innovative "solutions for fecal, urine, and menstrual management systems to be used in the crew's launch and entry suits over a continuous duration of up to 144 hours." The announcement garnered some five thousand submissions, from which twenty-one finalists were selected. Two of the three eventual winners offered innovative designs for garments, while the third person maintained that laparoscopic surgical techniques were the answer. In a moment of levity about the whole business, NASA added that among the competition's winners were the forty-six currently active astronauts, "who are very relieved."[1]

This competition was more than a playful public relations event. In fact, the problem of waste management has been a central part of the space age from the very beginning. Biological waste is the inevitable flipside of nutrition, and while on Earth the two processes are naturally connected via ecological cycles, in space neither one can be taken for granted. Andy Weir's 2014 novel *The Martian* describes vividly the intimate connection between nutrition and excretion, or food and waste. Weir's astronaut hero Mark Watney is accidentally left behind on Mars and faces the challenge of surviving roughly four years until the next mission is expected to land. Watney has four hundred days' worth of prepacked meals, which are tasty but finite, but also twelve valuable potatoes that were intended for the team's Thanksgiving celebration on Mars. Conveniently, the Mars habitat generates ideal growing conditions for these potatoes, and in a memorable scene in the book (and later film), Watney creates soil to grow his potatoes from a handful of Martian dirt by adding water, bacteria from samples of earth (from Earth), and, finally, his own packaged feces and urine as fertilizers. "My asshole is doing as much to keep me alive as my brain" becomes not only the hero's greatest one-liner but also a succinct description of a core element of twentieth-century space research.[2]

Weir's novel hinged on a necessary reality of humans going beyond the moon, which has been the subject of decades of research in space biology and medicine as well as environmental and systems engineering. From the beginning, both the American and Soviet space programs aimed at going to Mars and beyond. That ambition implied that people would have to spend months or more likely years in space travel, no matter what engine they used or how sleek their rockets were. (One science fiction story of 1940, the first to envisage a generation star ship, imagined a six-hundred-year trip into space.) During these journeys, the inhabitants of the spacecraft would consume considerable amounts of oxygen, food, and water, while at the same time producing proportional quantities of carbon dioxide and various bodily excretions. The two-thousand-person generation ship that Kim Stanley Robinson imagined in his novel *Aurora* (2015) would produce about 315 kilograms of feces per day (150 grams per person per day), or a discomforting 18 million kilograms of human excrement over the course of its 159-year voyage to a neighboring star.[3] Even if this kind of space travel was far beyond the space programs' actual aspirations, the problem reached alarming dimensions fairly quickly when contemplating three-year voyages to Mars, a goal of both Americans and Soviets.

While the problem was simple and obvious, the solution proved difficult to attain. Since it was practically impossible to bring all the necessary food and air for the long journeys from Earth, it would have to be produced on the way. To produce food and air, the space travelers would have to take care of their liquid, gaseous, and solid excretions. The conclusion was that the material cycles had to be closed, as they were on Earth. It was the operating assumption for more than sixty years, in both the United States and the Soviet Union, that long-term life support in space required astronauts to use biological waste to grow their own consumables in meticulously controlled artificial environments. The "waste-processing" component of any life-support system must be equivalent to the "food-processing" component and indeed the "crew itself" (see fig. I.1). How space travelers attempted to make this work is, in part, the subject of this book, which tells the story of how scientists and engineers of the space programs

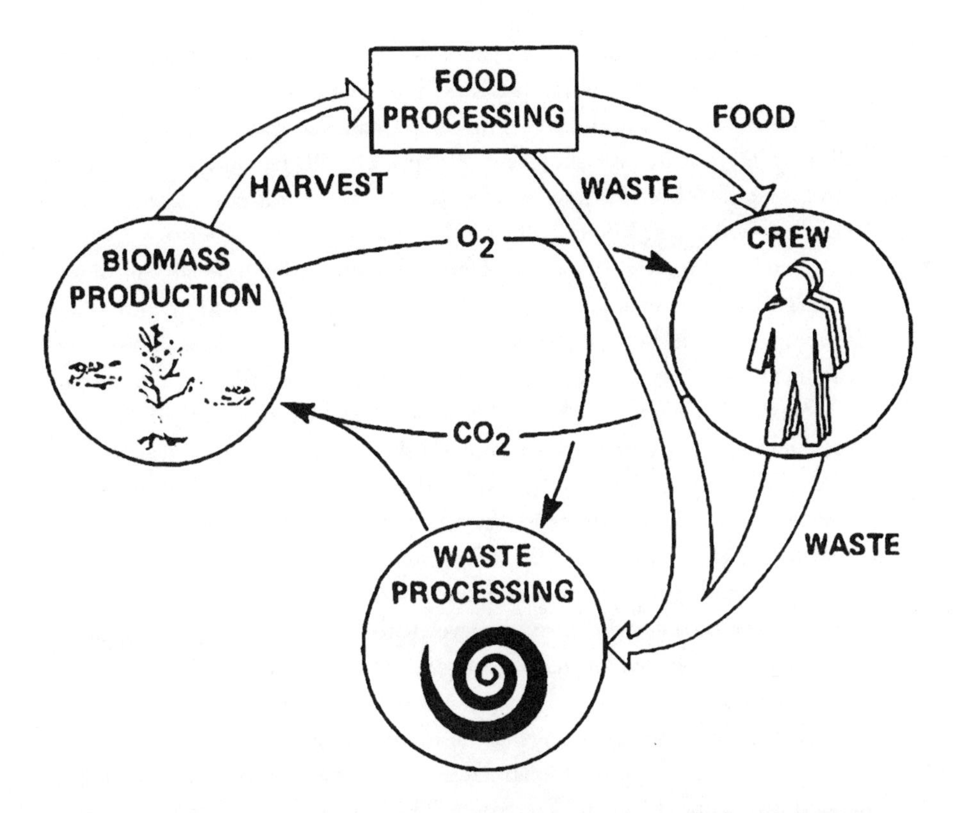

Fig. I.1. "Material cycling within a bioregenerative life support system." From R. D. MacElroy and James Bredt, "Current Concepts and Future Directions of CELSS," in Robert D. MacElroy, David T. Smernoff, and Harold P. Klein, *Controlled Ecological Life Support System: Life Support Systems in Space Travel* (NASA Conference Publication 2378; Washington, D.C.: NASA, 1985), 2.

converted visionary thinking into material reality. The chapters cover the period from the late 1950s until the early 1990s—starting with modest attempts to replace storage devices on board with regenerative algae systems, and ending with the ambitious large-scale projects to replicate whole ecosystems on Earth, such as the BIOS-3 project in the Soviet Union and the Lunar-Mars Life Support Test Project and Biosphere 2 complex in the United States.

Building appropriate environments to sustain humans in space was not considered an unrealistic goal, merely an extremely difficult one. Ever since the late 1950s, the space age sought the science and technology of what we would now call sustainable resource management. It produced a wealth of ecological knowledge of the functioning of closed environments. It also confirmed the insight that a regenerative life-support apparatus was the main factor that determined how long (and how comfortably) anyone could live in space. But the insight itself was not yet the solution. Even today, the quest to build a completely closed artificial environment remains unfulfilled. As the Poop Challenge indicates, space agencies still struggle to create even short-term systems. One of the more successful and promising efforts over the past decades has been the European Space Agency's MELiSSA system, the Micro-Ecological Life Support System Alternative, which used microbes to recycle air, water, and food to astronauts for deep-space missions. In a 2017 interview, its leader, Christophe Lasseur, succinctly explained that the European Space Agency's program and agenda was "to characterize all processes in as much detail as possible as a first step to recreating it, based on the knowledge we acquire." This was the rationale from the start of the space age. In order to make life-support systems work, a comprehensive investigation of all life processes and their interconnection was required and thus NASA and other agencies always promoted large programs in the life sciences and ecology.[4] Yet decades later, even after the first trial runs of the MELiSSA system have been completed, the space age is still far from a fully functional closed life-support system. For all our knowledge of life processes on the molecular level, without understanding the complex interrelations of the components of ecological systems, this task is challenging.

These massive research programs in the construction of artificial environments have not received anything like the attention they deserve. Most historical and popular accounts of the human adventure in space invariably have looked upward and ignored its earthier aspects. This was no coincidence. The space race was a spectacular turn during the heyday of the Cold War, that struggle between distinct ideological visions of society.[5] After 1945 the Soviet Union and the United States vied for international hegemony by becoming military-industrial behemoths with large bureaucracies and secret worlds of security, missiles, and surveillance. By the mid-1950s, however, this rivalry had produced a global power stalemate. The conflict thus evolved into a war of images, rhetoric, and above all grand technological displays, and the space race became its most extraordinary show.[6] Over the next thirty years, it became a competition to see who put the first animal into orbit, then the first person, a landing on the Moon, and eventually who laid the foundations of a permanent human presence in space. "The goals of the [space] program are not scientific goals; they are political," clearly declared the first chairman of the Atomic Energy Commission, David Lilienthal, in 1963. Political leaders on both sides worked to make sure that the correct message was received both by their own people as well as the global community. What was perceived as "correct," however, was very different. It was a vital component of the space race that both sides competed to legitimate their version of events. As historian Asif Siddiqi notes, Russians still see the launch of Sputnik and the later

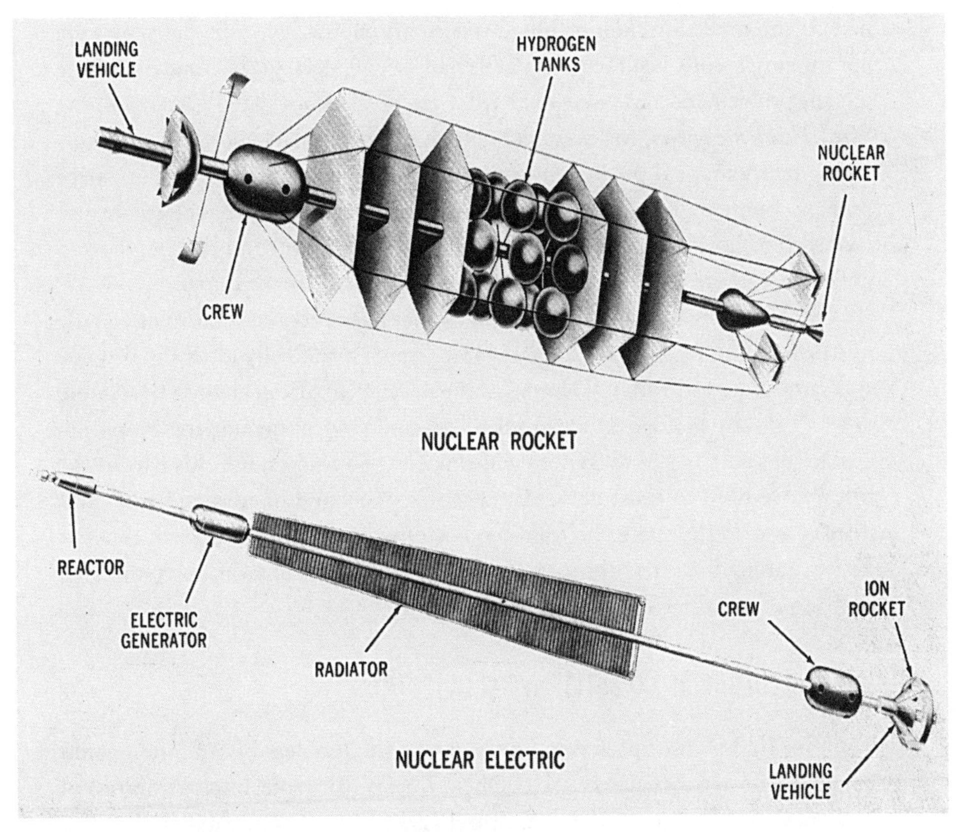

Fig. I.2. "Space Ship Structures." From National Aeronautics and Space Administration, *The Challenge of Space Exploration: A Technical Introduction to Space* (Washington, D.C.: NASA, 1959), 25.

orbit of Yuri Gagarin as the greatest breakthroughs of the century, while the landing on the moon is perceived as only a minor consequence of their earlier triumphs. Americans, of course, see those events in reverse: the landing on the moon is the real feat, while the Soviet accomplishments were just a prelude.[7]

The space race served its political purposes so well because it was a media spectacle covered exhaustively in print and televised everywhere. Countless TV shows, coffee table books, novels, movies, and popular histories parade the splendor of the space programs, with gleaming rockets and movie-star cosmonauts and astronauts next to their Jackie Kennedy–lookalike wives. Their predominant historical narrative stresses stories of engines, heroes, and power, and thus replicates earlier patterns of equally masculine automobiles where horsepower and design had set the tone.[8] Speculation about spacecraft engines was rife, but they were not the entire story, just the most visible part (see fig. I.2). At the same time more socially problematic parts were rendered invisible. The book (and Oscar-winning film) *Hidden Figures*, for example, tells the story of black women computers whose mathematical prowess proved crucial throughout the American space program but remained largely uncelebrated. The working realities of gender and race with a short-lived female astronaut training program and the long-lasting use of black women computers went against how NASA wanted to be perceived.[9]

Our book emphasizes other aspects of the space program that went equally undebated in public but were equally important, specifically ideas about waste and systems to deal with it. Decades in the making, artificial environments were developed—by biologists, social scientists, and environmental engineers—to enable the recycling of waste into consumables. To understand this part of the history, we need to resist being dazzled by rockets and handsome heroes and instead look at the space age from the bottom up, rather than the top down.[10] To live among the stars, ironically, has always meant solving the down-to-earth problem of sustainable waste management.

NASA'S ECOLOGICAL EXPERTISE AVANT LA LETTRE

The first flights into space were so short that the problem of sanitation could be put off. In 1964 Michael G. Del Duca, chief of the biotechnology branch of NASA's Office of Advanced Research and Technology, recounted in his keynote address to a major conference how, "in early manned space flight, much attention was given to space feeding and nutrition, but the problem of waste handling was eliminated by eliminating elimination."[11] As Del Duca noted, adult diapers sufficed for flights of a day or two, while a condom attached to a tube and complemented by a plastic fecal bag permitted flights of up to two weeks. Bags and diapers sufficed for nearly two decades. Incremental progress came in the 1980s, when a space commode was installed in the first space shuttle (employed from 1981 until 2011), and then later in the International Space Station (ISS) since 1998. The commode remains "the most important piece of equipment to master," as astronaut Scott Kelly put it when he arrived at the ISS for a full-year stay in 2015.[12] All these devices work through collection and storage—and the latter to an excessive extent. Solid waste has been returned to Earth throughout humanity's adventure in space (allegedly now sitting in jars on shelves at NASA and elsewhere), while carbon dioxide has been filtered and, until the ISS, urine and condensate expelled. The techniques were considered "well understood, relatively compact, low maintenance," and perfectly sufficient for short-duration

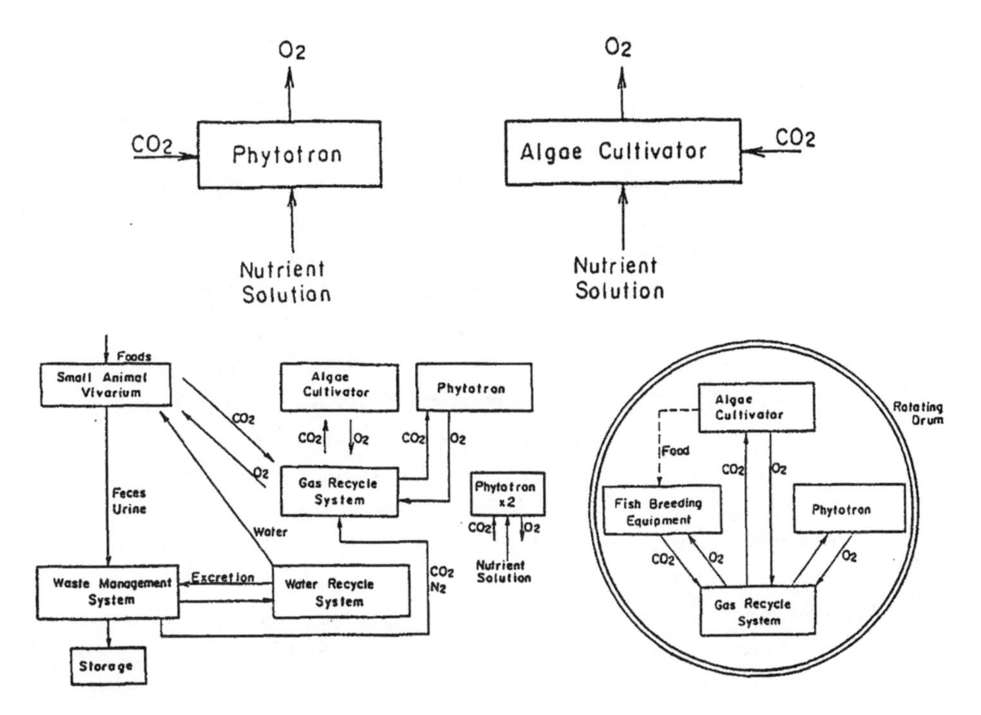

FIG. I.3. "Experimental Concepts." From Haruhiko Ohya, Tairo Oshima, and Keiji Nitta, "Survey of CELSS Concepts and Preliminary Research in Japan," in Robert D. MacElroy, David T. Smernoff, and Harold P. Klein, *Controlled Ecological Life Support System: Life Support Systems in Space Travel* (NASA Conference Publication 2378; Washington, D.C.: NASA, 1985), 14.

flights.[13] But everybody knew that other solutions had to be found for more ambitious endeavors.

From the start, people like Del Duca took it for granted that regenerative systems had to be developed that returned excreted waste to the crew as sustenance. Algae systems were among the space age's favorite alternatives. Norman Bowman's description of the uses of algae for food and waste recycling appeared as early as 1953 in the *Journal of the British Interplanetary Society*, and the notion that this might be the way to go has never faded since. Consequently, many people thought that the space programs had to carefully study algae cultures and acquire profound ecological knowledge if they wanted to conquer space. It was not enough to think of spacecraft as vehicles that were able to travel long distances; one had to envisage them as moving ecosystems. In fact, when NASA planned a new space station in 1985, one discussant, Sharon Skolnick, noted that they were "sending up a new planet, actually a microcosm."[14] Within such microcosms would be interconnected waste management and gas recycling and water recycling systems connecting the crew to their environment through plant growing phytotrons, animal vivariums, fish breeding equipment, and algae cultivators (see fig. I.3).

The ambition on both sides of the Iron Curtain was to create self-contained environments in which people could live for years, perhaps even decades or centuries. Those environments were envisioned as little copies of Earth traveling through space. Indeed, it was an analogy drawn quite early on. Already in 1962, the American sanitation engineers William Oswald and Clarence Golueke noted that their small-scale algae/waste system for space was really "a miniature version of the grand scale terrestrial ecological system of which we are a part." And the image persisted over the next decades: sixteen years later, in 1978, Robert MacElroy and Maurice Averner, two pioneers of NASA's Controlled Environment Life Support System Project, maintained that "an isolated system capable of supporting human life . . . bears a resemblance to the whole terrestrial ecosystem."[15] It was unanimously agreed, throughout the period under study here, that in order to build the former one needed to understand the latter: in order to survive in space one had to investigate how humans survived on Earth, and then try and replicate the conditions—just as Lasseur described the approach of project MELiSSA.

There was, however, no romanticism here. American and Soviet engineers and life scientists did not attempt to reconstitute a fanciful "Edenic Nature" where people would live without labor or concern through recycling waste. Rather, engineering and scientific professionals demanded that to live in space would necessitate a distinct realism for a space station or generation ship to support life. In direct and earthy ways, one popular writer Tom Allen noted in 1965 that any long-term spacecraft would be a "cloacal dwelling place."[16] Such arable biological metaphors contrasted the sterilized futurist "visioneers" of the late 1960s and 1970s, who imagined that humanity's colonization of space would be the opportunity to establish new social arrangements alongside a new architecture. Even in the 2010s the realism of NASA challenged the disinfected assumptions of observers: ethnographer Valerie Olson "sat for months" in meetings listening to life scientists and engineers talk about "waste" "conjoined" with humans and machines. She defined it as "crazy-weird work in cultural terms" but observed that it was "rather matter-of-fact in environmental systems technical terms," betraying her own assumptions about the reality of living in space and the work of NASA.[17]

As this book describes, within the mainstream space programs the entire spacecraft was conceived as an earthy techno-ecological system. The physical challenge was to build such "regenerative" techno-ecological systems into machines that regenerated the fundamental biological and chemical resources of life processes in systematic feedback loops, befitting the cybernetic thinking of the time. A "cybernetic method of thinking," one Soviet space biologist noted in 1965, helped to "find an analogy between the biological phenomenon and the processes which take place in engineering devices" to reveal biological knowledge. This was, to use historian Cyrus Mody's term, square science and technology. Around 1970 a new "groovy science" of the counterculture emerged in distinction to the militarized and mainframe academic culture.[18] However, argued Mody, these divergent directions left out the square middle, scientists and technologists with little sympathy for the counterculture but eager to attack problems such as public transportation, housing, water, and in the case of life scientists at NASA, waste.

A startling consequence of the space age's square and practical cybernetic thinking was that the human occupants became but one set of components among others. Like algae or machines, humans too processed energy and nutrients and were as replaceable as others. The American science writer Mitchell Sharpe wrote in 1969 that men and machines "cannot be thought of as separate entities" in space. Cosmonaut Yevgeny Shepelev, the first person to live in an enclosed artificial test environment, said in 1965 that "man here is an object to be safeguarded only insofar as he can ensure the normal functioning of the other links in the system." Such views of the comparative values of man and machines challenge the still commonly held assumptions of the entire space age and the centrality of the human to life in space. Olson was equally surprised when she observed in her ethnography that today's NASA as an organization treats the astronaut as an "ecosystemic" rather than a biological entity, but we argue that this has always been the case.[19]

Much of the pertinent work was done at within the Life Sciences Division of NASA's Ames Research Center in Palo Alto, California. Another section of that division was exobiology, which evolved over time into the field now known as astrobiology.[20] The task of exobiology to imagine and pursue ideas about and detection of extraterrestrial life represented only a small fraction of NASA's efforts to create the conditions in a vehicle to move life into space. Within the Life Sciences Division and elsewhere, scientists and engineers at Ames sought to provide values for the rates of cycles, density of organisms, and ranges of tolerances to environmental parameters in order to define the limits of life on Earth and in space, or what is generally known as habitability. Everything about the environment that supports life had to be questioned, since nothing was self-evident in space. Could human beings function in zero gravity, or would they be reduced to spasms of nausea? What about the impact of radiation above the protective atmosphere? How would humans deal with not knowing which way was "down" (in space, there is no up or down), or would their inner ears just keep them spinning? In 1963 NASA's director of its bioscience programs, Orr Reynolds, asked about "the possible occurrence of subtle cellular effects" from a human's mere presence in space, "which might alter basic biological processes." To answer such biological and medical questions, the United States and the Soviet Union cooperated and shared information at international symposia beginning as early as 1962. But forty years later many of the same questions are still being investigated. In 2015 astronaut Scott Kelly and cosmonaut Misha Kornienko spent a year aboard the ISS for test purposes because, as Kelly later explained, "very little is known about what happens [to the human body] after month six" in space. A crucial component of the experiment was that Kelly's twin brother on Earth provided a control to measure space's effects on a body.[21]

Answers to the immediate questions came rapidly—yes, humans were able to function in zero gravity; and, yes, some of them were, in fact, constantly nauseous—but the larger questions were not as easily answered. Public interest in these questions, apparently, was high. A 1963 *Voice of America* broadcast, for example, assured the audience that humans performed normally in space and noted the "reliability of algae" even in the intense magnetic fields. To assemble a life-support system, however, was going to take more than men and algae. As the interviewee said, a critical complication was that the network of environmental factors was so complex it resisted experiment. In 1963 botanist Colin Pittendrigh was called to testify before the U.S. House Committee on Science and Astronautics. He detailed at length that many plant physiologists, space scientists, and engineers had offered ideas about the "environment" of space. Replicating such an environment was still little short of impossible: "Clean questions and clean answers are going to be difficult [because] when we put organisms into space at present and detect deleterious effects, we are simply unable to disentangle the many variables that exist there and decide which has been responsible," Pittendrigh explained. Like many scientists of his generation, Pittendrigh tried to deal with these problems by using controlled facilities, but success remained limited. Even decades later, the challenge has not disappeared: nature remains understood as a highly interconnected system within all space programs, and understanding its dynamics and functions is still regarded as extremely difficult. "Whatever part you describe first, the description cannot be complete without the knowledge of the other parts," the leaders of both the Soviet and American life-support projects concluded in 2003.[22]

It has long been assumed that the space programs pulled such concepts about the environment from emergent ideas in ecology. Prominently, historian Robert

Poole argued that the environmental consciousness of the 1970s could be traced back to the space race, notably as pictures of the Earth as a pale blue dot became widely reproduced. Spaceship Earth—the globally famous metaphor that was coined and distributed by Barbara Ward and Buckminster Fuller and served as a conceptual vision of the Earth as a spaceship traveling through the universe—is often claimed to have raised ecological concerns also among space scientists. As should already be apparent, this narrative requires substantial revision. Rather than ecologists, the American and Soviet space programs relied on plant and animal physiologists, microbiologists, nutritionists, and environmental and sanitary engineers. That ecologists played little role bewildered many: University of Washington ecologist Frieda Taub pointed out as early as 1974 that "sealed ecological systems have been surprisingly unexplored as ecological tools." Taub found that academic ecologists were largely unaware of a vast pool of research from the space sciences into microcosms, regulating mechanisms, and bioregenerative life-support systems, on which the United States had spent an estimated $30 million up to 1966. While it is true that the important ecologists Howard and Eugene Odum proposed a bioregenerative system for the American space program, it was only a theoretical project and had little impact at NASA—indeed, it was prematurely terminated in favor of more realistic endeavors that had been going on for some time already.[23]

The pragmatic environmental sensibility that came out of Stewart Brand's *Whole Earth Catalog* from 1968 to 1972 or paraded at the first Earth Day on April 22, 1970, was much in evidence within 1960s NASA, an organization that was the very antithesis of Brand's small-scale communities and individualized technologies. The practical efforts of NASA to build miniature systems in the 1960s was a parallel, applied strand to the emerging concepts of Spaceship Earth, biosphere, and carrying capacity, which, as historian Sabine Höhler demonstrated, came to dominate the environmentalist movement of the 1970s and 1980s. While ecology became regarded as the "subversive science" because of the work of Rachel Carson and Barry Commoner, the transformation from old-style conservationism to modern environmentalism saw environmentalists of the 1970s push the (allegedly) new and pressing insight that within Spaceship Earth all resources were finite, and thus a sustainable economy and recycling systems were imperative if humans were to survive much longer.[24] However, this was not an entirely new thought. Already during the 1960s, NASA had built life-support systems that fully relied on solar power and had bioregenerative cycles implemented—which could have been, but were not, the foundation for the new and fashionable countercultural ecovillages. On Cape Cod, Massachusetts, John Todd and the countercultural New Alchemist Institute built an eco-village that produced food by means of sustainable agriculture and aquaculture, apparently without any knowledge of the existing research into these very issues by NASA.[25] The Living Machine system built in the Findhorn eco-village in 2003 uses a series of connected barrels with plants and algae as a sewage system, while nobody seems to have been aware that it was almost identical in design to a system built by Boeing aerospace for NASA in 1963.[26]

We argue that NASA, the epitome of the military-industrial complex, and the Soviet state, were equally unexpected sources of ecological awareness, heralding a holistic conception of life in space and on Earth remarkably earlier than the emerging countercultures that stimulated an environmental movement. The modern ideas of environmentalism have blossomed since 1970, exactly the moment when NASA largely ended its first major research and development effort into waste recycling. Those within the life sciences of the space age were keenly aware of how limited and valuable resources were in a closed system. In fact, in many ways the solutions of the space programs were more radical than those of the countercultures. In 1971 a group of notable Soviet scientists announced their conclusion that life support in a closed system "consists of the members of the system eating each other's metabolites." More startling still, *Time* magazine's science editor was quoted by famed author Arthur C. Clarke as suspecting that in order to ensure complete closure of an artificial environment for long-duration space travel, "cannibalism would be compulsory among interstellar travelers."[27]

Such confronting opinions were public, as was much of the research on closed-environment life-support systems by the military-industrial complex on behalf of NASA. Similarly, the Soviet Union widely advertised their progress toward living in space. In the sharp criticisms of the complex in the late 1960s, however, the knowledge that the highly technocratic space program had developed workable ecological systems to recycle air, food, and waste was largely forgotten (like that shown in fig. I.4). By the 1980s NASA stood very much in the vanguard of ecological thinking. Working on the proposed space station *Freedom* in August 1983, Jesco von Puttkamer from NASA's Office of Space Flight noted that the current United States faced the same situation as any future orbiting habitat, namely a growing population swelling against an "originally pristine environment," "increasing amounts of waste into a finite containment," and the "effluents of industrialization."[28] In other words, the American designers of the future space station were fully aware of the environmental situation of the United States itself, and they used that troubled legacy in their work. It was clear, however, that the results of NASA's work challenged the self-conception of Western society at the time. An astronaut on board a

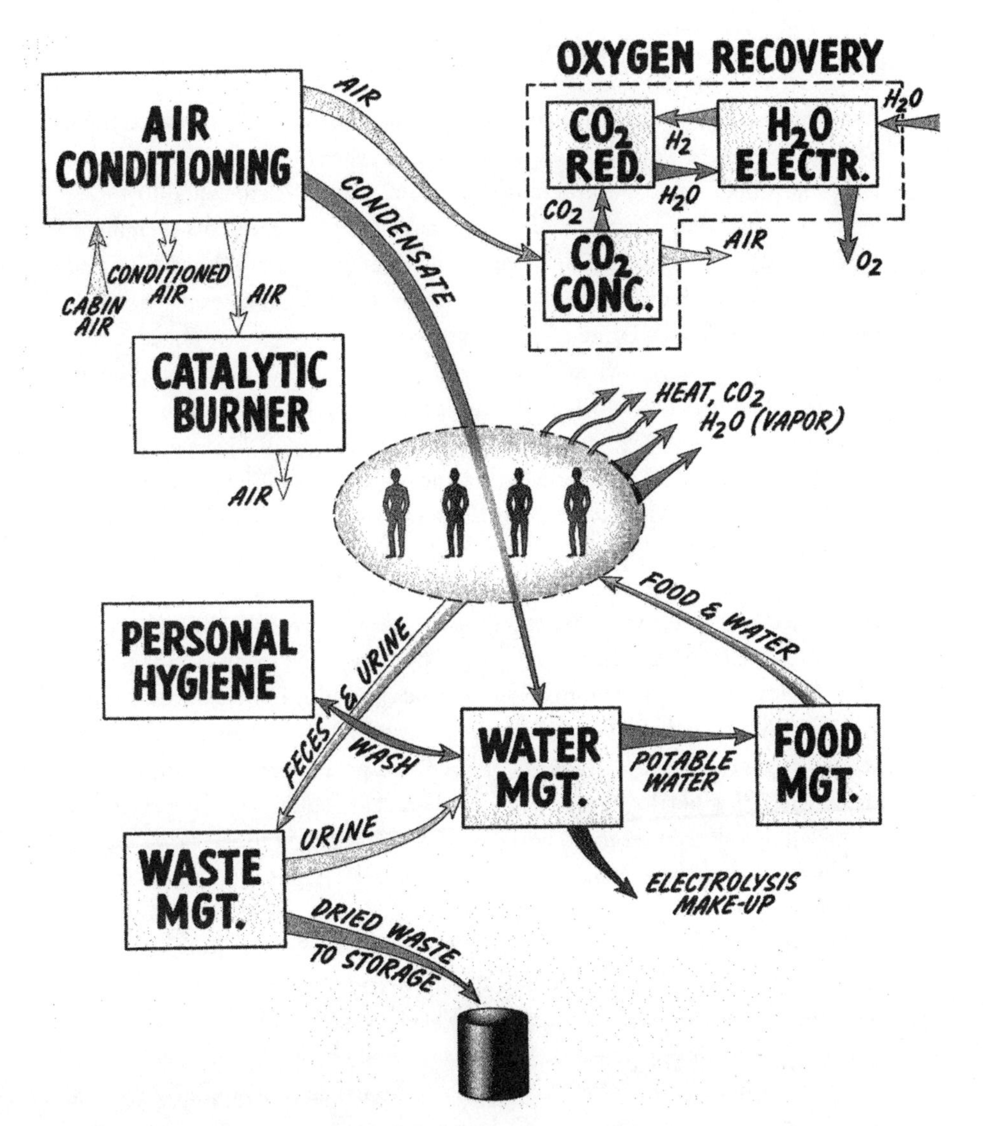

Fig I.4. "Flow diagram of the Integrated Life Support System (ILSS) regenerative process." From Judd R. Wilkins, "Man, His Environment and Microbiological Problems of Long-Term Space Flight," NASA TMX 60422, 1967, 26.

spaceship breathing air recycled from carbon dioxide, drinking water recycled from urine, and eating algae recycled from excrement was shocking to the era's social sensibilities, even though it greatly appealed to the engineer's notions of efficiency and elegance.

WASTE AND ITS MANAGEMENT IN HEAVEN AND ON EARTH

As should be clear by now, the history of artificial environments involves talking a lot about waste, especially biological waste. By centering waste in our history, our project takes up historian Donald Worster's call for excremental histories. Worster claims that historians have not paid enough attention to excrement, bowdlerizing environmental history by focusing on where food comes from and not where the waste goes.[29] In a significant social shift in human history, since the 1950s people in the United States (and elsewhere) got used to the idea that "waste" could just be thrown away. Few comparable studies of waste complement a wealth of studies of Cold War consumption, even as the emergence of the "throw-away" society arguably remains the era's most pernicious legacy.[30] The new lifestyle created all manner of waste, including air pollution, refuses, solid, liquid, and gaseous wastes, from municipal and industrial sources. Their ensuing problems were acknowledged only in the late 1960s.[31] A memorable portrayal of the era's attitude to waste was John Kenneth Galbraith's *The Affluent Society*, in which packaged food and new cars contrast littered streets and polluted streams. Galbraith's take on "American genius" can be seen in the iconic period drama *Mad Men*, when the protagonist Don Draper and his family go for a picnic to a park. At the end of the meal, his wife Betty first checks the cleanliness of her children's hands before shaking off from the picnic blanket the unwanted containers, drink bottles and caps, paper plates and napkins. Leaving the trash all strewn on the grass, everybody walks unconcerned back to the car.[32]

While people have only halfheartedly talked and thought about trash, another form of waste, namely sewage, has been rendered silent, odorless, and invisible throughout the twentieth century.[33] Social historians like Donald Reid, Joel Tarr, and Martin Melosi have described how American and European cities confronted epidemics in the late nineteenth century by beginning to build water and sewage systems, but mostly as patchworks of local facilities. In consequence, those cities pushed their problem literally "downstream"—to the detriment of neighboring towns and villages.[34] At the scale of a house, too, the common household water toilet linked to a sewage system was above all "designed to hide" and carry away human excreta. As we shall see, the containment of the same substances in a fecal bag by early NASA astronauts as a stop-gap solution or in the space commode by later ones operated similarly to flushing (if less efficient in eliminating odor). Both containment and removal rendered the substances invisible, at least to the observing public. Rendered equally invisible in the historiography, the topic of sanitation has not found its way into serious descriptions of the space age or the Cold War life sciences. If it is mentioned at all, the topic is used to entertain children.[35] In colloquial presentations, of course, astronauts have regaled their audiences with stories about defecation in space, consciously playing with the notion that this more than anything demythologizes the glory of space exploration. The larger story of waste management in space, however, is little talked about: how bodily wastes—urine, feces, and sweat—became nutrients (see fig. I.4).

Social inhibition contributed to the fact that the problem of sanitation on Earth or in space was for a long time trivialized and underestimated in the face of political reality. Plant physiologists and agricultural experts had known for a long time that waste forms half of the material cycles in nature: whenever and wherever water, air, and nutrients were absorbed and processed, the remainder was excreted. In nature, it reentered the cycle through fermentation by microorganisms, and nineteenth-century chemists and agriculturalists excitedly discovered that guano (and other forms of biological waste) could be used as fertilizer. But a century later sanitation professionals in the United States noted that their field was critically short on trained manpower and that its research institutes were woefully underfunded, although the growth of new suburbs obviously increased the problem of sewage. In October 1957, when Sputnik had just started to orbit, the leading sanitary engineering department in the United States at the University of California, Berkeley, warned about "the multitude of problems of environmental sanitation" that accompanied "the explosive growth of California's cities and their attendant industry and agriculture on limited land and water resources." In a bold move, President Lyndon Johnson took up the cause of making "America the beautiful" as part of his Great Society vision. With bills addressing air and water pollution, wilderness preservation, and solid-waste disposal, Johnson accomplished an impressive legislative record by the mid-1960s. Major sanitation systems, however, took decades to improve: as late as the 1980s, raw, untreated sewage still flowed directly out of Boston into Boston Harbor, out of New York City into the East River, and out of Los Angeles into the ocean.[36]

The problem of sanitation was also viewed as a problem of pollution and removal rather than of recycling and reuse. Awash in new chemicals with doubtful effects on human health, the early environmental movement of the 1960s focused on specific substances marked as pollutants. The careless disposal of sewage was criticized alongside the spread of pesticides, nuclear waste, and synthetic detergents, without discriminating between substances of biological and nonbiological origin. These clearly were campaigns to reduce the contamination of fields, streams, and cities rather than movements inspired by a rising ecological awareness.[37] That ecological awareness, however, as our book reveals, was very much in place at the space programs—without, of course, being combined with global, politically motivated, anticapitalist aspirations.

Waste management became the subject of well-funded science and engineering projects within the space programs, building on the idea that if sanitary systems could be perfected to fully return waste products as food, water, and air, "humanity could truly and biologically tear ourselves away from the Earth's biosphere." Engineers and scientists looking toward space began to conceive of waste in ways that obviously differed from how it was understood on Earth. The most significant of those new understandings emerged in the late 1950s when space biologists tried to use the waste of one organism as a nutrient for another. They saw success with animals in closed systems, witnessing how mice and algae were able to support each other for weeks on end while closed ecospheres of shrimp and algae in water lived on for years. Ever since, the space sciences have had a radically different notion of waste than that of society at large; there really is no waste in space. As Wendell Mendell of NASA's Mission Science and Technology office said during the planning for Space Station Freedom in 1991, "the term 'waste' becomes an oxymoron [in a closed-loop life-support system] because every atom contributing to organic chemistry is precious." That "waste" might become "nutrients" was by then obvious to NASA's life scientists and environmental engineers but remained a tough concept to accept for even the most liberal environmentalists of the 1970s and 1980s: as Dorian Sagan noted, it "is difficult to wax poetic about medical waste, chlorofluorocarbons, and carbon dioxide. Yet . . . excrement, garbage, trash—all of the most rancid and marginal parts of our anatomy—are one day transmuted into parrots, wine grapes, magnolia trees."[38]

THE SCOPE OF THIS BOOK

Our book claims that research into artificial environments, including the development of regenerative sanitation systems, was an important part of the space age that has been neglected. Furthermore, this book claims that in pursuing this research, space programs on both sides of the Iron Curtain generated profound ecological knowledge about the functioning of interrelated, complex systems. From the early 1960s onward, American and Soviet life scientists and environmental engineers were remarkably informed about sustainable resource management. As the NASA life science division's collection of papers and reports from the Soviet Union's project display, the American and Soviet space programs exchanged considerable information as they worked toward the goal of a permanent habitat in Earth orbit or journey to Mars. Both sides of the Cold War were fully aware of the fact that they had to accommodate all parts of human existence into their life-support systems, and both started to think of human waste, in the sense of all bodily excrements, as being an indispensable component of material cycles. American and Soviet scientists and engineers concerned themselves not only with chrome surfaces, power engines, and combustibles but also with algae cultures, the nutritious value of quail and potatoes, and the composition of feces. And, quite radically for the context, those same scientists and engineers realized that their knowledge about how to live in space could be projected onto Earth as well. From these claims, we suggest that insights from attempts to lead a fully sustainable existence in space are useful to inform more recent attempts to secure our future existence on the planet in the face of the disastrous implications of climate change and other manmade catastrophes.

The story of the creation of life-support system to live in space spans both sides of the Iron Curtain. Our knowledge of the Soviet program is based primarily on sources from NASA, which collected and translated Soviet reports, articles, scientific publications, and popular pieces coming out of all aspects of the Soviet space program.[39] These documents offer initial insights. The challenge of language and other barriers precludes a complete description of the vital Soviet contribution to this story, but we eagerly look forward to future work on the subject.

Chapters 1 and 2 turn to the first twenty years of the space age, from 1957 to 1977, and reveal the major engineering and biological milestones to achieve an artificial self-sustaining environment, complete with material and energy recycling. In chapter 1, we describe how both the American and the Soviet space programs aimed from the start to go to Mars and beyond. The moon landing only was originally proposed by NASA as a stopover on the way to larger and longer missions. As we know, however, once the lunar landings were achieved, American political priorities shifted, and NASA's budget was cut immediately thereafter. Chapter 2 investigates in more detail how scientists and engineers addressed the challenge of sustainable life-support systems in these years—including a particularly instructive and well-documented case: the so-called Algatron, which was developed by sanitary engineers from Berkeley.

Chapters 3 and 4 describe the next twenty years of the space age, from 1977 to 1997. That period saw the first attempts to build a space station, which also—and necessarily—included life-support systems. Longer-term experimentation was needed to prepare these ambitious projects. Chapter 3 describes how the Soviet Union's early *Salyut* and later *Mir* space stations took advantage of the elaborate BIOS-3 facility at Krasnoyarsk, Siberia. In this research station, experts from Moscow's Institute of Biomedical Problems collaborated with the biologists from the Institute of Biophysics to develop appropriate environments. They tested complex life-support systems that accommodated several dozen organisms including wheat, algae, sweet potatoes, and *Homo sapiens*, eventually locked in for yearlong experiments. Chapter 4 begins with the American *Skylab* station as an experimental orbiting habitat, then continues to describe the planning of space station *Freedom*, which was supposed to be the U.S. response to the Soviet successes of the 1970s. By the 1990s, NASA believed itself to be almost there, with a bioregenerative life-support system within reach and a full-scale trial system experimentally proven. Many of its components are now on board the ISS, but complete recycling of material remains elusive. Unexpectedly, as explored in chapter 5, one of the most complete bioregenerative life-support systems came in the form of the Biosphere 2 in Arizona. The two crewed missions that lived inside the closed structure between 1991 and 1994 came closest to experiencing the full range of physiological and psychological pressures of the kinds of long-term space journeys the space age had long dreamed of.

Ultimately, in their attempts to build systems to live in space or on Mars, scientists and engineers gained crucial insights into how humanity may continue to live on Earth. Over the last twenty years, it was a question that haunted the creators of life-support systems. "What kind of knowledge must humanity attain in order to rationally govern the biosphere?" asked Iosef Gitelson and Robert MacElroy, the former leaders of the Soviet and American efforts in the 1990s.[40] There was at least one thing that the quest for life support taught the protagonists of the space age: embracing waste is part of life.

1

WHEN AMERICA AIMED BEYOND THE MOON

The lunar project is not the whole space program.

—Senator Stuart Symington, 1964

IT WAS 1963, AND AMERICA SEEMED TO BE LOSING THE SPACE RACE. THE LAUNCH of Sputnik, the first artificial satellite, by the Soviet Union in 1957 had deeply shocked the American public. In these years, everyone linked superior status with top-notch technology. As the popular magazine the *Saturday Evening Post* observed in 1969, both superpowers saw "their destinies as depending on technological achievement."[1] While Sputnik beeped into orbit, the assumption of American superiority appeared doubtful, perhaps even false. Desperate to claw back the technological lead, Congressman James G. Fulton of Pennsylvania famously asked the head of the new U.S. space agency (NASA) in 1958, "How much money would you need to . . . make us even with Russia? . . . I want to be firstest with the mostest in space, and I just don't want to wait for years."[2] The answer, of course, was that it would be years, and cost fortunes.

A rapid succession of dramatic firsts in space by the Soviet Union between 1957 and 1963 continued to embarrass the United States. In November 1957, just a month after Sputnik, America's space leaders learned that the dog Laika had allegedly lived for seven days aboard Sputnik 2 thanks to a functional life-support system. This turned an initial stoic concern into near panic. George Low, then in charge of planning manned spaceflight at NASA, wrote in his 1959 report that while both the American and the Soviets had launched animals into the upper atmosphere, the Soviet program appeared about a year ahead of NASA's. Laika's "air regeneration system" was a feat "we [the Americans] have not attempted," Low noted. Moreover, he lamented, "We will not be in a position to duplicate that achievement until July 1960, or one year after the Russians did it."[3] As it turned out, Low was a little too pessimistic. The United States managed to launch a squirrel monkey named Gordo in December 1958. He was equipped with a rudimentary survival system for a flight of just fifteen minutes. While "Gordo was provided with neither food nor water," he did have "oxygen from a tank" and "carbon dioxide absorbed by pellets of baralyme." However, "temperature control was only partially achieved by insulating layers of metal foil and fiber glass," and as was true of astronauts for many years afterwards, "the waste management system consisted of clothing the monkey in a diaper." Sadly, Gordo died upon reentry from parachute failure. In view of this situation, Low still predicted that the American presumption of technological superiority would erode further: "One might also speculate that when the Soviets accomplish manned orbital flight, they will put two men into orbit, as opposed to our one. Their entire history of biological flights in space is based on having two animals in each payload. They most certainly have the weight-lifting capacity for a two-man capsule. And the propaganda value of flying two men, with the knowledge that we cannot do this for many years to come, would be a great one."[4] It must have been cold comfort that when the Soviet Union did indeed successfully orbit the first man in space in 1961, Yuri Gagarin, he flew alone.

In hindsight, it is easy to underestimate how severe and frightening the Soviet challenge was at the time. As their rockets blew up on national television, all the Americans could do was move the goal posts: while visiting the Moscow exhibition in 1959 vice president Richard Nixon famously attempted to change the conversation from rockets to washing machines during the Great Kitchen Debate with Nikita Khrushchev. The loss of security through technology created a widespread loss of confidence in the American way of life. One former presidential candidate, Adlai Stevenson, pointedly asked, "With the supermarket as our temple and the singing commercial as our litany, are we likely to fire the world with an irresistible vision of America's exalted purposes and inspiring way of life?"[5]

Americans feared their nation's scientific leadership slipping away barely a decade after the atomic bomb supposedly permanently assured it. One response was to initiate a crash educational program to increase the numbers of trained scientists and engineers and to overhaul the entire science curriculum.[6] Another famous response was President John F. Kennedy's January 1962 announcement that the United States would land a man on the moon and return him to Earth by the end of the decade. Given the state of the American space program, Kennedy's claim was full of hubris, made with a keen eye toward the upcoming midterm elections. In line with this political context, the rest of Kennedy's speech outlined plans for foreign aid, disarmament, and closer ties with U.S. allies. But speeches did not immediately translate into achievements; indeed, the next year saw yet another Soviet first, notably the orbit of the first woman cosmonaut, Valentina Tereshkova. As historian Margaret Weitekamp noted, with the early cancelation of the Lovelace Woman in Space Program in 1962, the United States did not respond to the political challenge of Tereshkova's flight and did not achieve that inclusive feat until Sally Ride joined the early space shuttle mission STS-7 in 1983.[7] Watching the symbolic scoreboard, NASA's director of Biotechnology and Human Research, Eugene Konecci, unambiguously called out how badly the American's lagged: while Soviet cosmonauts had already logged 380 hours of orbital flight, their American rivals had chalked up only 53 hours and 26 minutes. Two years later, in 1964, the Soviet Union achieved what Low had already anticipated: they put two men into orbit aboard the spacecraft *Voskhod I.* Forced to acknowledge the Soviet milestone, NASA administrator James Webb sourly called the Soviet multi-manned mission a "significant space accomplishment." Still smarting as late as 1967, the American Congressional Committee on Science and Astronautics acknowledged in hindsight that "it was a long road for us, and we did not match the 1957 Soviet records for total weight in orbit in any U.S. payload until perhaps the Saturn I flight of 1964."[8]

After 1964 America clawed back some ground in the space race. In the years between 1964 and 1968, the components of what would become the Apollo program materialized. Rapid progress culminated with the Apollo 11 landing of Neil Armstrong and Buzz Aldrin on the moon in July 1969, but the preparatory missions were also highly significant. The Apollo 8 spacecraft first orbited the moon while Apollo 9 tested a new configuration with a command module (which would stay in lunar orbit) joined to a lunar module (which would land on the moon and return). Watching any of the numerous documentaries on the Apollo missions today, one might conclude that once the Eagle touched down in 1969, the space race was won and done. At the time, however, it was commonly understood that "the lunar project is not the whole space program," as the NASA publicity office reported a senator saying in February 1964. Lunar missions were not seen by NASA as an end in themselves but rather as a series of trials to work out the problems of being in space.[9] Indeed, since the beginning of the space age NASA had looked beyond the moon to space stations, interplanetary flights, and Mars missions. And as this chapter describes, that is where the story of life-support systems really started to matter.

FIRST STEPS TO BUILDING A SPACE STATION

A space station had been an integral part of NASA's thinking long before Kennedy's lunar announcement. NASA's Long Range Plan of 1959 slotted its first flights to establish a "permanent space station" in the period 1965–1967, while the "manned flight to the moon" was expected only "beyond 1970." In one of its earliest public documents of the same year, *The Challenge of Space Exploration*, NASA illustrated the expected growth of space vehicles (fig. 1.1). From an initial 2,200-pound payload (365,000 pounds of thrust) rocket required for a one-man, one-day initial mission, NASA envisioned continuously larger rockets needed for two-man, two-week missions to the moon, up to the 150,000-pound payload needing 6 million pounds of thrust for the rocket that would insert a permanent space station into orbit, which was the ultimate goal.[10]

Doing everything to attain this goal in time was to become the agenda of NASA's two centers of research. On the East Coast, there was the Langley Research Center in Hampton, Virginia, founded in 1917, which was matched on the West Coast by the Ames Research Center, founded in 1939 in Mountain View, California—right in the heart of today's Silicon Valley. Both were originally sites of aeronautical research, initiated by the National Advisory Committee on Aeronautics but transferred to NASA after 1958. Famously, Ames possessed the United States' largest wind tunnel, which had been part of the development of most American aircraft from propeller planes to the space shuttle.[11] But with the opening of the space age, NASA required expertise not only on streamlined rocket shapes but also on bioscience. From 1959 onward, researchers at both Ames and Langley began to study the behavioral, medical, and biological facets of space flight. This included investigations into metabolism, nutrition, weightlessness, and artificial environments. As the space race heated up, research in these directions intensified with the parallel institutional expansion of the agency in the early 1960s.

The first of NASA's human space flight programs was Project Mercury, running through 1958 to 1963. The goal was to successfully bring humans up into space and safely return them to Earth. Limited life-support systems were immediately indispensable even for this initial, comparatively modest enterprise (fig. 1.2). Langley became the primary site for developing an appropriate space capsule and a space suit. One of its early outcomes was the environmental control system, which supplied oxygen to the space capsule, controlled cabin temperature, maintained cabin pressure, and automatically "removed metabolic products" within a "zero g condition and in a high g acceleration."[12] In a much-needed triumph, the environmental control system functioned perfectly on Gordon

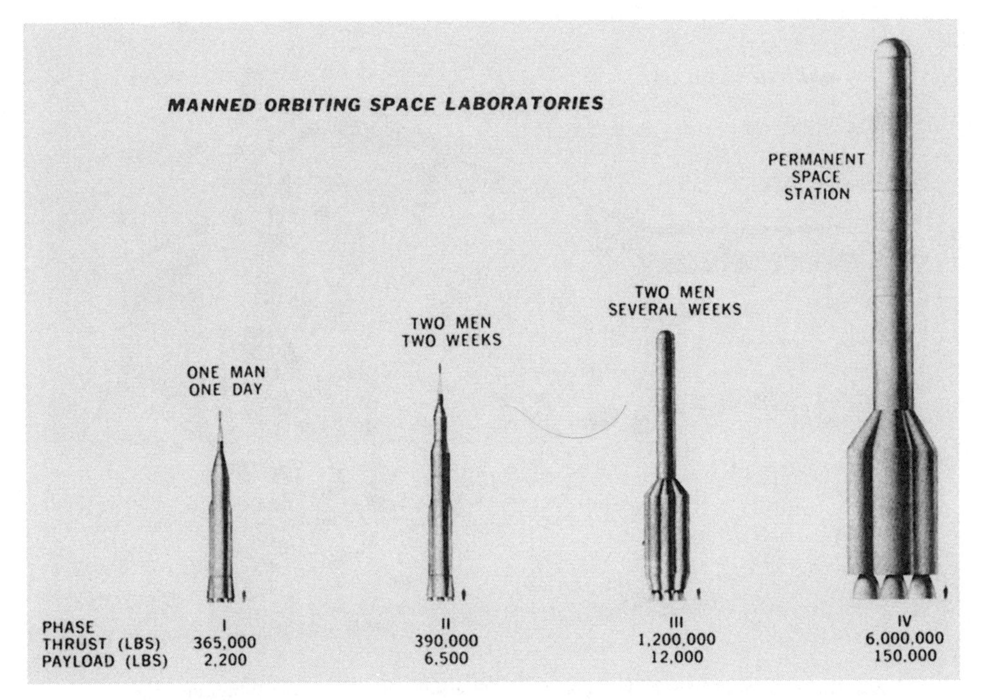

FIG. 1.1. "Expected growth of 'Manned Orbiting Space Laboratories.'" From National Aeronautics and Space Administration, *The Challenge of Space Exploration: A Technical Introduction to Space* (Washington, D.C.: NASA, 1959), 42.

Cooper's thirty-five-hour orbital flight in May 1963. Cooper hardly had time to settle into a bodily routine during this flight, but it was long enough for NASA to assess his physiological response to zero gravity, as well as sleeping, eating, and excreting in space.

By the time Cooper was back on Earth, multiple groups were already feverously working to create larger environments to sustain life in space for longer than a day. Jesse Orlansky, an analyst for the Institute of Defense Analyses, one of the new technical consultancy firms sprouting up on the beltway of Washington, D.C., surveyed the efforts within NASA and the United States Air Force in the field of bioastronautics: "a relatively new term [that] includes medical, physiological and psychological studies related to airborne and space flight."[13] Orlansky counted five separate projects directly related to studying life support in space among three major institutions. They included the "food and waste management" project and the "O_2 reclamation studies" at the Manned Spaceflight Center in Houston and, at Langley, the $900,000 "breadboard life

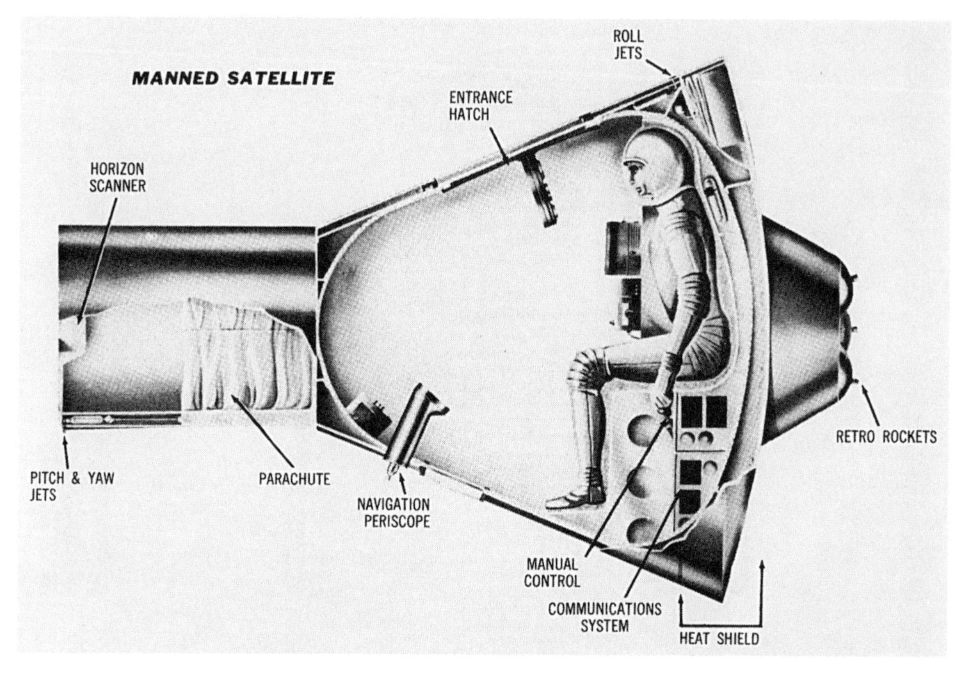

FIG. 1.2. "Manned Satellite." From National Aeronautics and Space Administration, *The Challenge of Space Exploration: A Technical Introduction to Space* (Washington, D.C.: NASA, 1959), 41.

support system" that was intended to keep four men alive for a whole year in a 2,600 cubic feet cabin and "capable of recycling all water (except from feces)." According to Orlansky, there was unanimous agreement that progress in these fields was crucially important if humans were to go into space or even live there. He also drew attention to the "paradox of the present [1963] DoD-NASA relation." Orlansky noted that while NASA had manned missions it was really the air force that possessed "the largest bioastronautics training and research facility in the world," the Aerospace Medical Division in San Antonio with a staff of more than a thousand and an annual budget of $35 million.[14] This situation would continue for years to come and was only one of the indicators that space flight was by no means a predominantly, let alone purely, civilian project.

With the Mercury program successfully completed, NASA looked ahead to the longer multi-manned Gemini and Apollo missions. For even the two- and three-person crews, the rudimentary life-support features of the Mercury capsules needed to be improved and the research devoted to their development had to be broadened. NASA's Biotechnology Research Advisory Committee

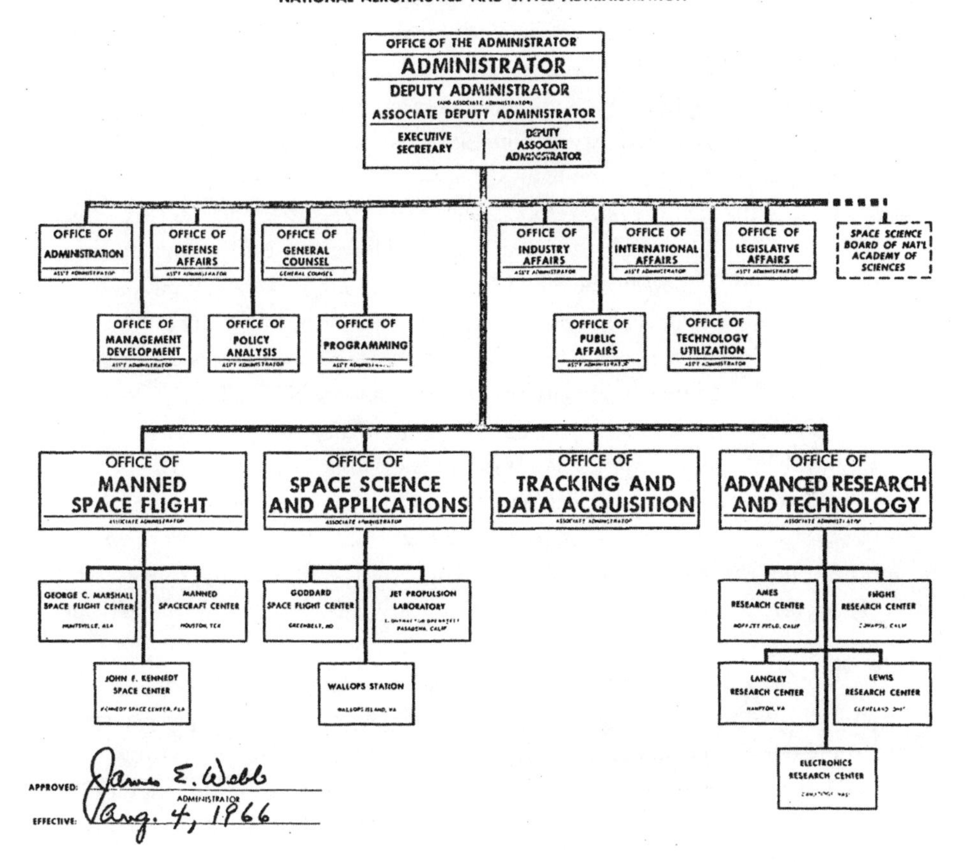

FIG. 1.3. NASA organizational chart, August 1966. Available at https://history.nasa.gov/org charts/39HQorg66-8-4.pdf.

pointed out that Langley had a "highly sophisticated engineering knowledge" of life-support systems but lacked a "deeper technical capability in the biological, physiological, and psychological disciplines."[15]

NASA thus pressed the Life Sciences Directorate at Ames into service under the leadership of Harold Klein. Organizationally, the Life Sciences Directorate was one of four branches of scientific and technological coverage at Ames, alongside aeronautics, astronautics, and development (see fig. 1.3). The life sciences section, in turn, was divided into three divisions, namely exobiology, environmental biology, and biotechnology, each one occupying around twenty research staff. The Langley Research Center, by way of comparison, had around thirty scientists and engineers concerned with biological and medical studies, who after 1963 were moved into a new Life Support Technology Laboratory. In hindsight, exobiology, the forerunner to today's astrobiology, has achieved the highest recognition, far beyond its institutional origins at NASA. Its main interest became identifying organic molecules that might indicate extraterrestrial life, but it also investigated the conditions of habitability of potential alien planets and what forms "life" itself might take out in the wide universe.[16]

Klein welcomed the new responsibility for Ames' life sciences department. Klein had done his graduate work at Brandeis University and took up the post of the head of the new Exobiology Division in January 1963 before taking over the Life Sciences Directorate. He consistently stressed the importance of the life sciences in NASA's mission. By the late 1960s, as the life sciences diminished alongside manned missions in favor of probe missions at NASA, Klein headed up the Mars Viking lander team to try and determine the existence of life on Mars. Even as NASA geared up for the flight to the Moon in the mid-1960s, Klein stressed that it was not sufficient for NASA to be concerned with the immediate tasks required by the Apollo program but had to think about the missions ahead. NASA thought that way too: between 1961 and 1966, NASA awarded about sixty contracts to major aerospace companies to study the methods and technologies for a journey to Mars.[17]

Amid the preparatory work on a space station, NASA's Ames Research Center built a fifty-foot centrifuge between 1962 and 1963 to study pilot performance "required for a lunar or planetary flight." This centrifuge, however, was also used to produce forces to establish "design criteria for future rotating space stations." The image of a space station rotating through the universe, in order to simulate gravity for its inhabitants, captivated the enthusiasm of both NASA and the wider public. The science-fiction writer Arthur C. Clarke had already imagined a rotating space station in his 1951 novel, *The Sands of Mars*. In 1968 it reappeared in Clarke's most famous novel, *2001: A Space Odyssey*, where he declared that "the space station, or permanent manned orbiting structure, may be regarded as the next step beyond the brief extra-atmospheric excursions which opened the Space Age." Shortly thereafter, theatergoers saw a rotating space station on screen in Stanley Kubrick's film version of *2001*.[18]

Clarke's and Kubrick's vision of the future was quite to the point from NASA's perspective. On the evening before Neil Armstrong and Buzz Aldrin walked on the moon, journalist and novelist Norman Mailer interviewed the head of the Manned Space Program, George Mueller. Mueller used this opportunity to regale Mailer with his own vision of a spacefaring future far beyond the moon. This future comprised a multitude of space shuttles and space stations in orbit, nuclear power plants on the moon to supply water and oxygen, and "a livable

FIG. 1.4. Drawing of a lunar farm. Used under a Creative Commons license from the Space Sciences Institute. Available at http://ssi.org/space-art/ssi-sample-slides/.

atmosphere within an enclosed space," where space pioneers "would even grow plants."[19] For Mueller and so many others in these years, the prospect of realizing earlier utopian visions of lunar farm lay just over the horizon (see fig. 1.4).

ENGINEERING THE BIOLOGICAL CONQUEST OF SPACE

There was a lot to be done to make these visions come true, however. At the same time as rockets had to be made more powerful and reliable, NASA put a huge amount of time and effort into figuring out the physiology of diet, digestion, and excretion in space as well as the dynamic interaction of these processes in the human organism and its environment. The role of the life sciences in the space program was not a salve to the consciences of rocket engineers as ecology arguably was to the American nuclear regimes.[20] In fact, the life sciences were central to NASA's goal of taking men into space. Throughout the 1960s, Ames, Langley, the United States Air Force, and aerospace contractors such as Lockheed, Boeing, and General Electric took on the task of developing experiments to understand how to live in orbit. These significant institutions, known primarily as doyens of the military-industrial complex, also sought ways to regulate and stabilize the interconnected cycles of nutrients, organisms, and environments.[21] Life in inhospitable environments was a major part of the science and engineering of the early space age: at the Goddard Memorial Lecture in Washington in March 1966, Lorne Proctor spoke at length about the central importance to NASA of life support in space. In "hostile environments" like the moon or Mars, Proctor said, "oxygen, carbon dioxide and inert gases are the major problems, and the most efficient mixtures of these compounds within a suitably closed ecological system has been and still is receiving much attention."[22]

Unfortunately, most of the interesting and significant work in the life sciences was subsequently left out of histories of NASA, even its own histories. According to the official version of 1983 by David Compton and Charles D. Benson, concerns about artificial environments were minimal in NASA's first decade, since space adventures remained limited to around two weeks. Only when NASA graduated to the prospect of living in space in the 1970s, Compton and Benson claim, was advanced sanitation technology developed including the space toilet.[23] Edwin Hartman, who wrote the first history of Ames, made the same point. Privileging the beginning of exobiology over any other habitability studies, Hartman describes how NASA's life science division had begun work on exobiology only around 1970 by investigating how human physiology reacted to a space environment. They exposed rats to yearlong 2.5 g, 3.5 g, and 4.7 g environments in cages mounted on the ends of a centrifuge, yet "closed ecological systems, involving the recycling of all wastes," Hartman said, were allegedly still only "being considered."[24] (This work on rats was also forgotten. During negotiations with Elon Musk around 2002, Mars Society founder Robert Zubrin wanted to send mice spinning in a capsule to test their ability to reproduce, evidently not aware that this had been done decades earlier.[25]) Revising the narrative that substantial work on the physiological, ecological, and social aspects of space flight was only conducted from the 1970s onwards in closed-environment systems is the goal of this chapter.

A major study by the Space Sciences Board of the National Academy of Sciences concluded in 1965 that the "exploration of Mars—motivated by biological questions—does indeed merit the highest scientific priority." From its earliest years, NASA had devoted "a great deal of attention," as the organization itself put it, to closed ecological systems to underpin both a permanent space station and ultimately a mission to Mars. Moreover, that work was highly regarded at the time: designing and building a complete life-support system offered "a variety of interesting and complicated challenges to engineering ingenuity," by no means inferior to constructing the rocket itself.[26] In what became a common refrain,

NASA's Research Advisory Committee on Biotechnology and Human Research noted in 1965 that "long duration manned space flight will require recycling or regenerative life support systems." At the same time, the committee proudly announced that first steps toward this goal had been completed: "A system capable of recycling water and oxygen has been assembled at Langley. It will support four men for 90 days without resupply." The ability to sustain teams of four to six men out of direct Earth contact for three months, NASA thought, was the minimal requirement of any future system. "Long duration," however, meant pushing even further than that. Among others, NASA's microbiologists sought experimental runs of "a year or longer" if they were to assess the epidemiological risks of bacteria and viruses spreading through the envisaged space station.[27]

A recycling system was essential for any mission longer than a few weeks. Supplying sufficient consume-and-discard food and water to a group of astronauts on a three-year mission to Mars amounted to enormous quantities and, hence, weight. Weight remains one of the absolute and unforgiving constraints of every space program. "Weight is the Enemy," stipulates Andy Weir's hero astronaut in *The Martian*. Everything taken on a space mission must be essential because everything adds weight, and weight adds cost to any spacecraft launch. Even in the twenty-first century, the transport of goods from Earth to the International Space Station comes at a cost of approximately $10,000 per kilogram; that is, lifting a single glass of water costs about $3,000—not including the glass.[28] Of course, NASA engineers knew that a recycling system also weighed something, and, importantly, it demanded power to operate. Consequently, there was a breakeven point at which a recycling system paid for itself in weight savings. Langley concluded as early as July 1963 that "because reclamation systems require large amounts of power, for flights up to 20 to 30 day's duration it is more economical to supply air and water and dump them overboard." Beyond space ventures of a month, however, things looked fundamentally different.[29] A NASA diagram drafted in 1965 clearly indicated that the initial difference between regenerative and non-regenerative systems was the startup penalty from the equipment itself and its power drain (see fig. 1.5). But as the intersection point specified, the weight penalties beyond weeks- to months-long missions made stored expendable supplies increasingly unrealistic.

A similar diagram appeared six years later in a completely different context: in G. Dennis Cooke's 1971 essay, "Ecology of Space Travel," a chapter in Eugene Odum's seminal textbook, *Fundamentals of Ecology*. It was from this latter diagram that the idea of recycling in a closed system, which would permit long-term sustainability over expendable stored supplies, gained public attention. Generations of students would study Cooke's diagram of the breakeven calculation

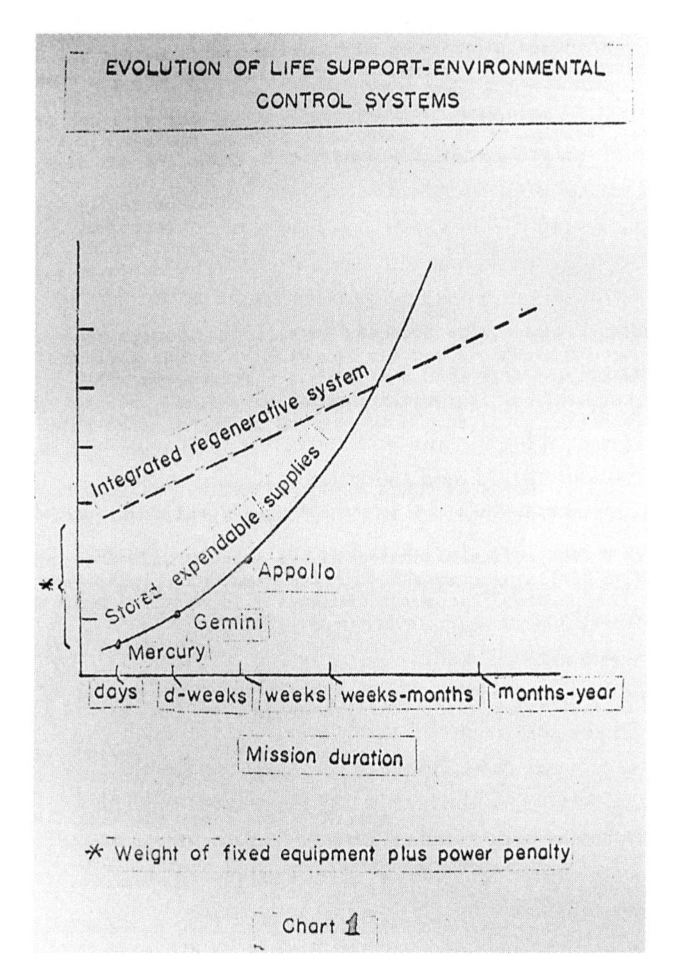

Fig. 1.5. "Evolution of Life Support Environmental Systems," December 6–7, 1965. Folder "Biotech & Human Res. Cte. Mtg 12/6–7/65," Box 1, Series 36: Life Sciences Directorate, 1963–67, RG255, NARA (acc. no. 255-93-022).

between non-regenerative versus partial and total regenerative systems.[30] Conceptually, Cooke's diagram displayed exactly what NASA had already concluded several years earlier. Yet, because the ecologists' version dropped (or did not know) the scale information on the time versus weight axis, one could not place a value on the breakeven point—which, in effect, made the diagram and the intersecting lines practically useless. In contradiction to another well-rehearsed standard narrative, the new discipline of ecology played little part in the early attempts to create closed environments. Historians have connected ideas about living in space to the rising science of ecology during the 1970s, but on closer inspection there is little evidence that the science of ecology had any influence on the engineering of life-support systems within and without NASA.[31]

WORKING THE "MAN-MACHINE" AND "MAN-SYSTEM" PROBLEM

It is instructive to look a little closer at how the scientists and engineers at NASA actually tried to work the problem of life support. NASA's Biotechnology and Human Research program was officially formed in September 1962. Its task was to develop adequate systems for survival in, as they still said, "the aerospace environment." Primarily, this was "human research," measuring the various environmental stresses to be faced by astronauts. The organization of the research program indicates how NASA approached the challenge. Within the topic of "man," for example, it investigated how the "body systems" performed by looking at "environmental physiology," which could be measured via "bio-instrumentation monitoring." At the same time, under a heading "Advanced Concepts," NASA already looked to create "human analogs (cybernetics)" and pursue the field of "bionics." All this was only the first stage of an expansive feedback research program. As psychologist, and human factors specialist, Stanley Deutsch, the executive secretary of the Advisory Committee on Biotechnology and Human Research, noted in a 1964 internal report, "the human, man-machine, and man system requirements must be determined through research prior to the design of any aeronautical or space system." The program tested the performance of "life support systems," "protective systems," and "man-machine control" systems and developed "environmental, H_2O, food, waste control" and "information displays controls." In view of "man-system" interactions, the program sought answers to questions around mission performance reliability and "maintainability." Remarkably, the whole assemblage was framed as an "integral ecological system."[32]

Over the next years, NASA funded a host of similar research projects with the aim to "derive design data through experimentation, mathematical modeling, and simulation for the integration of man into the design of advanced aerospace systems." In the cybernetic spirit of the times, such systems were intended "to enable men and machines to function as an integral unit."[33] NASA was clearly convinced that new technological environments enabled space flight. The challenge, however, was to fit the human in as part of the artificially created environmental systems. As cosmonaut Yevgeny Shepelev, the first person to seal themself inside an artificial environment, put it, man had to become in this system "a constituent, one of its function links." Space suits for short excursions by an astronaut outside their spacecraft, EVAs (extravehicular activities), were an immediate miniature case of the larger problem of how to design the human-machine-environment complex. Man-machine control, bionics, and bio-monitoring: in hindsight, this all sounds curiously similar to the vision of space advertised in *Star Trek*, the first season of which appeared on televisions in 1966. *Star Trek*'s early episodes dealt regularly with the meaning of being human in a machine dominated world and with the survival on alien worlds and in space.[34]

Among the tasks of the earliest crewed missions were critical assessments to judge how far an astronaut could be considered "a reliable operating portion of the man-spacecraft," as the director of the office of Biotechnology and Human Research, Eugene Konecci, phrased it in 1963.[35] NASA saw the astronaut and the spacecraft as two components of a larger complex system. The idea of the "man-spacecraft" was expanded over the next few years. Among the immediate objectives was that NASA needed to establish the "physiological and psychological requirements of humans in multi-manned systems utilized for missions of 30 days to 6 months," as well as the "human design requirements for supersonic transport, hypersonic transport, advanced X-15, lunar colony, [and] manned Mars." Notably, from the perspective of NASA engineers and life scientists, it was the human component that was bound to be the most troublesome. "We must recognize the capabilities and limitations of man," a Human Factors System summary of 1964 concluded, before adding more optimistically that "by catering to the human's needs for life and efficient performance, we will ensure that *he will not* be the weakest link in NASA's aerospace systems."[36] The idea that the human element was the most troublesome part of the forthcoming era of space exploration had long been mentioned. At the First International Symposium on Basic Environmental Problems of Man in Space in Paris in 1962, K. Steinbuch argued that the limitations of acceleration on a human body (no more than 15 g) compared to those "electronic systems [that] can stand accelerations from hundreds or even thousands of g," like those of temperature and radiation, made man not suitable for space in comparison to "automatons." Even some twenty-five years later, Yevgeny Shepelev would relay to the new crew of the Biosphere 2 in Arizona essentially the same message: "humans are the most unstable element in the ecosystem. Have courage."[37]

THE MILITARY'S MANNED ORBITING LABORATORY

One of the earliest attempts to engineer this array of biological ideas and speculations into a working space station was the Manned Orbital Laboratory (MOL). NASA's original conception of the MOL was a laboratory for experiments that could not be conducted on Earth. "There is no substitute for zero gravity studies in a manned orbiting laboratory with long term capability," NASA noted in 1963, as it listed out the MOL's expected research avenues. Before permanent space stations or interplanetary missions, NASA needed more information on human health and performance under conditions of weightlessness and artificial atmospheres. To that end, the MOL would demand the near constant monitoring of the health of the crew, which lived inside a vehicle that would trial "habitability designs" and the "interaction of astronauts with the life support systems."[38] The MOL aimed to put two men into space for periods of at least thirty days. As Edward Welsh remarked in his keynote to the Third International Symposium on Bioastronautics and Space Exploration in November 1964, this was at least double the "seven or perhaps even fourteen day" periods of the "lunar program" and as such "will help determine just how well man can perform specific missions in space." While Project Mercury was a "toe-hold" in space and Project Apollo offered a "foot-hold," Welsh thought the MOL offered to move the whole human into space.[39]

The rapidly expanding demands on Langley to work up the MOL overstretched an organization already struggling with the Gemini and Apollo programs. In September 1966, about a year after he moved from Langley's Environmental Physiology Branch to the Biotechnology and Human Research Division of Ames's Life Sciences Directorate, physician John Billingham complained that the summary statement of the manned space station had "too much emphasis on 'monitoring.'" As Billingham wrote to Walton Jones, the director of the division, "surely, we've had enough of this already, producing untold miles of EKG, temperature, blood pressure, and respiratory rate recordings, most of which is completely negative. . . . We need investigations and experiments and research now, and not blanket physicals to the nth degree."[40] The organization most keenly interested in those miles of data was the United States Air Force (USAF). Not only did it possess most of the medical research and monitoring facilities concerned with extreme environment flight, it also worked toward its own manned orbiting space station.

The USAF had a particular interest in the area of man-machine systems and the physiological functioning of a human under environmental stress. It cooperated closely with NASA and had already built test facilities to evaluate human health and performance under extreme environmental stresses like high altitude and high g-forces, as well as low oxygen and low temperature conditions. In 1966 Harold Klein received an invitation from the NASA-AF Coordination Committee to attend an annual meeting to coordinate "the life support and space suit programs" between NASA and the USAF. He also subsequently received copies of USAF work for the preceding year.[41] With the space program gaining success and respect, this odd situation appears to have come from the air force itself to ensure that they were properly remembered. The USAF unashamedly took credit for essentially the entire space program, maintaining in internal magazines throughout the 1960s that all the space programs' elements had grown directly

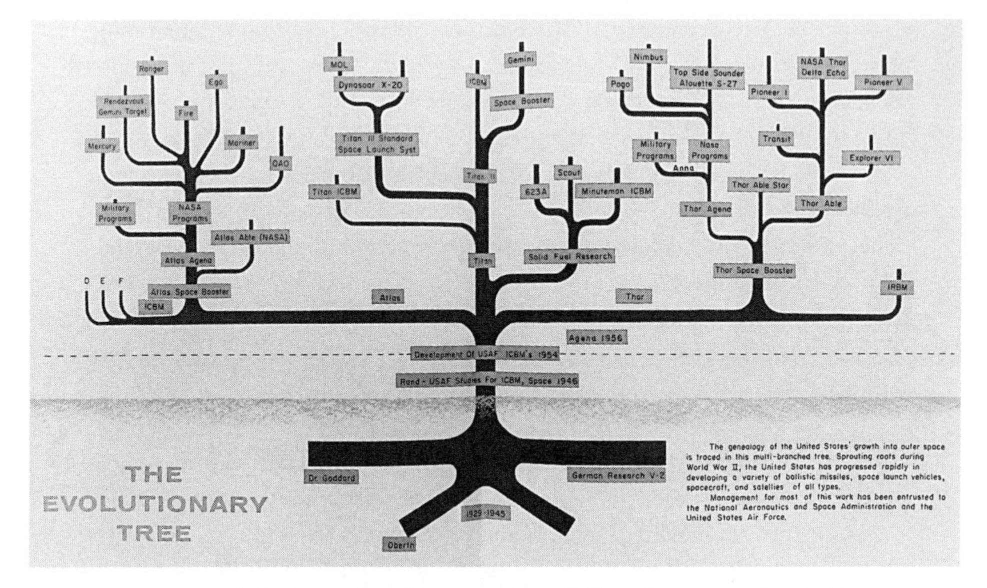

Fig. 1.6. Centerfold, GE Missile and Space Division, *Challenge* (Summer 1964): 25. Folder "MIT-Scientific Advisory Board-Manned Orbiting Laboratory," Box 37, H. Guyford Stever Papers, Gerald R. Ford Library.

from the air force's development of intercontinental ballistic missiles in 1954 (fig. 1.6).[42] While this was a great exaggeration, the military had, of course, never been far from America's space program, and its proximity had often benefited NASA. For a decade before 1957, the USAF tested not only new rockets but a whole series of biological samples including live monkeys and mice in what became known as space medicine.[43] In December 1965 Stanley Deutsch said that NASA had recommended research to evaluate "the capacity of humans to withstand or adapt to the rigorous demands of aerospace flight" because NASA urgently needed studies of human performance and response to "varied atmospheres, acceleration, [and] radiation." They all knew, of course, that another governmental body, the USAF, had already explored these topics; indeed, as Deutsch added, "wherever possible DOD [U.S. Department of Defense] research facilities are used" to save time and money.

Sometime around the summer of 1964, the air force and General Electric's Missile and Space Division took over development of the MOL from NASA's Langley Research Center—by that point, at the latest, it was no longer a secret that the American military was going into space.[44] According to the air force's own materials, the MOL, headed by air force Major General Ben I. Funk, explicitly looked to "determin[e] the military usefulness of man in space," notably the ability to "maintain and repair equipment," and "carry out scientific observations." But since the MOL was to be the size of a "small house trailer," part of the research remit was aimed at establishing "the minimum elements of personal comfort."[45] The USAF too needed not only a toilet but a whole sanitation system.

Arthur C. Clarke was much excited by the MOL proposal, which he saw as the first definite stage of establishing a permanent space station. Yet he was also concerned that "a veil of secrecy has descended (or ascended?) over the subject." At least one part of the MOL, however, became highly visible when the military announced on the front page of the *New York Times* for July 1, 1967, that Robert Henry Lawrence Jr. had been selected for the crew. Lawrence, a major in the USAF, skilled test pilot, and doctor of physical chemistry, thus was to become the first African American astronaut.[46] It was about time for space flight to become more inclusive: critics charged that the all-white face of the space program tarnished its national role as a unifier in increasingly turbulent years. Gil Scott-Heron's poem "Whitey on the Moon," the Southern Christian Leadership Conference protesting at Cape Canaveral, and Norman Mailer's account of the moon landings all cut to the heart of social tension cleaving American society as the 1960s ended.[47] Mailer saw how "putting two White men on the moon" served to celebrate "the most successful part of that White superstructure which had been strangling the possibilities of . . . Black people for years." But the immediate prospect of including a black astronaut in the space program evaporated when the MOL project was shelved in mid-1969.[48] It would not be until 1983 that an African American would finally get into space, with Guion Bluford on STS-8.

BUILDING A BIOREGENERATIVE BIOSPHERE INSIDE AND OUTSIDE NASA

As the MOL came and went, NASA continued the long process of forging the first space station potentially toward a Mars mission. It received strong support in this endeavor from the National Academy of Sciences (NAS). In a statement to the press in late 1964, the academy's Space Science Board declared that the prospect of exploring Mars "may well provide a most compelling justification for Apollo."[49] The statement had come in response to a request from NASA administrator James Webb, who had reached out to the chairman of the Space Science Board, Harry Hess, for a larger view on NASA's long-range plans. Webb was a former United States government undersecretary and corporate vice president who assumed the top job at NASA in 1961. That he asked the senior organization for American science about NASA's future plans spoke to the elevated levels at which such plans were being discussed. The official NAS statement was delivered on October 30, 1964, and emphasized that scientific priorities should guide NASA's agenda after 1970. To that end, the Space Science Board explicitly recommended "the exploration of the nearer planets as the most rewarding goal on which to focus national attention for the ten to fifteen years following manned lunar landing." The NAS Board concluded that the lunar mission would not be especially demanding (or instructive) beyond the mere continuation of life during launch and in a weightless environment. Rather a future Mars mission still required "the solution to difficult biomedical and bioengineering problems" to make NASA "ready for manned planetary exploration by 1985."[50]

No interplanetary mission could realistically succeed, the NAS Board acknowledged, until "the biomedical problems of long journeys are solved." In fact, "the ultimate goal," according to NASA's Space Station Requirements Steering Committee in 1966, was "to close all loops in order to provide a self-sustaining ecology." The NAS's and the committee's views were also fully in line with the policy of the Research Advisory Committee on Biotechnology and Human Research. As should be clear by now, a full range of NASA boards and committees were preoccupied with these topics at essentially the same time. While their views did not always coincide neatly, life and environmental sciences and engineering occupied the center of their bureaucratic jockeying for NASA's future. Quite explicitly, the Research Advisory Committee had underlined already in 1963 that in the period "1965–1975 more emphasis should be directed towards the planets leading to approximately even support for lunar and planetary exploration in the 1970–1985 period." The committee expected future interplanetary missions to follow on rapidly after the successful landing on the moon. To achieve that new goal, it asked for additional funding for 1967–1968 to pursue "very advanced biological closed cycle life support systems," including "CO_2 concentration, O_2 recovery, cardiovascular suit, and water, waste, food, and clothing management."[51]

First and foremost, this project was assigned to the Ames Research Center. Ames turned to its many contractors of the military-industrial complex, as well as its own in-house research groups. None of this was new but continued the NASA way of research and development. According to a January 1966 report by W. Hypes from Langley, NASA had already cooperated with private industry for about a decade to create equipment that could "regenerate useful materials from the waste products available in manned spacecraft." The outstanding issue, Hypes argued, was that the components had to be tested within a system because each was reliant on the "output of a previous component."[52] Cybernetic thinking met ecology at Ames.

Many of the usual suspects were also ready to meet this challenge, with the first to offer their services being General Electric's Missile and Space Division in 1963. Only one year later, Lockheed joined in and presented NASA a succinct "problem statement" of the critical issues. The company, of course, starred as the most significant aerospace contractor in the United States. It had already supplied much of the new air arsenal of the United States Air Force, and it would wheel out its famous SR-71 Blackbird strategic reconnaissance aircraft two years later, in 1966.[53] Lockheed's engineers calculated that the weight of food and water (the combination of which becomes human waste) quickly overwhelmed the payload capabilities of longer missions. Adding up the "metabolic wastes," including urine, utility water, expired carbon dioxide, and feces, the total number came to "over 100,000 lbs" for a ten-man, three-year mission even "without considering the weight of containers."[54] These values no doubt left an impression, considering that the Saturn V rocket used to launch the Apollo missions to the Moon could only insert a total weight of 104,000 pounds into a lunar orbit. In other words, without a recycling system the entire cargo weight of this monumental rocket would have to be apportioned just to lift the necessary food and water—and that was not going to work.

In the meantime, Boeing also built a life-support facility. Under the direction of R. H. Lowry from Boeing's Space Medicine Division, Boeing had been working on life-support systems since at least 1960. An early triumph for Boeing came when the company announced to the press that Lowry had himself lasted six hours inside a chamber breathing air excreted by living algae (see fig. 1.7). Later that year, Boeing bioscientist James Schubert spent fifty-six hours in the chamber.

By 1963 Boeing had built on this rudimentary beginning to produce the first full-scale life-support facility for NASA complete with a "waste disposal system [that] collects, processes and recovers water from body waste, cabin condensation, and sink drainage." It even had an early version of a "closed circuit shower" for a "crew member to bathe in." However, the arrangement was far from perfected: promoted to Boeing's chief of bioastronautics, Lowry, together with Eugene Konecci, captivated the International Astronautical Congress in 1963 with the gruesome tale of the first test run.[55] Unbearable nausea and "profuse perspiration" greatly affected all five crewmembers after only two days. On top of this, the toxins released by either the polyvinyl sleeve of the internal electrical cables or the synthetic rubber tubing (the precise source was unclear) rose to dangerous levels. Then came the mechanical failure of the waste reactor and the appearance of a "yellow, oily substance found in the humidity condensate." It is no surprise that the crew lasted only five days, and even this indicates an impressive willingness to persevere and suffer.[56] In hindsight, the trial seems irresponsibly rushed, with

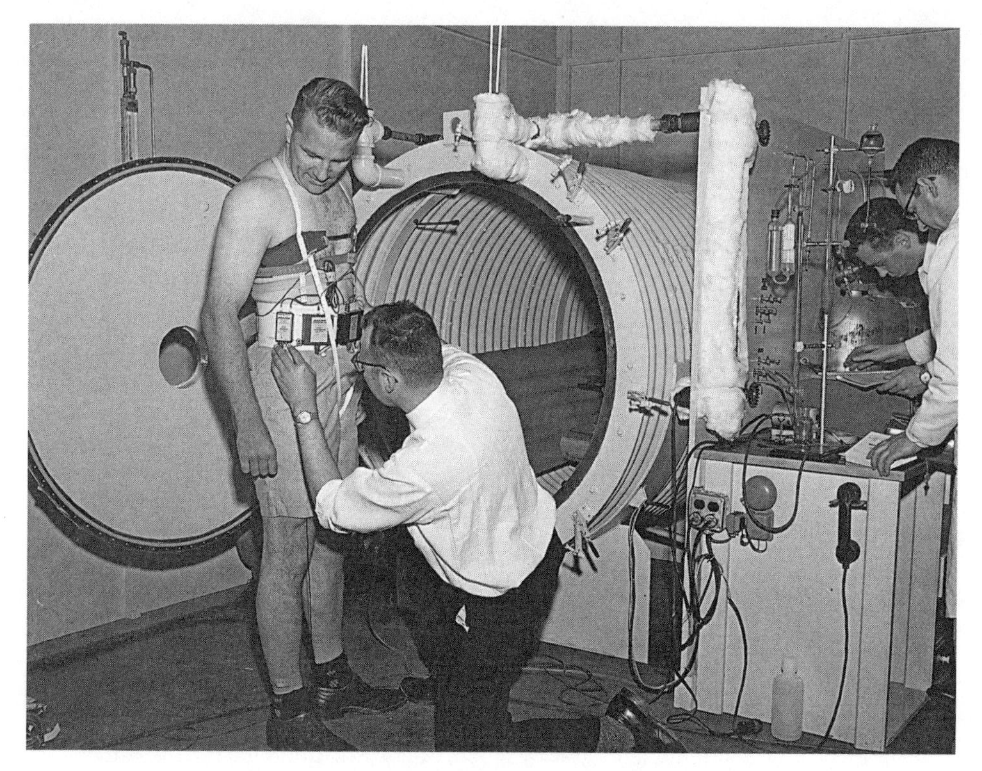

FIG. 1.7. R. H. Lowry preparing to enter the sealed algae chamber for six hours. Boeing Space Medicine News Release, May 11, 1960, P24816. Courtesy of Boeing Aircraft Co.

little forethought: all they knew was that a "nebulous" array of "toxic compounds" arose when a "combination of materials and equipment [were] integrated into a total system." Keenly aware of a gap in knowledge, NASA's Langley Research Center launched a major research project in 1966 to investigate what contaminants existed in such a closed environment, as well as their sources and their toxicity. As the assistant director for Life Sciences at NASA, Harold Klein, admitted in a meeting that year, "Virtually nothing is known about human production of contaminants" within closed environments.[57]

Shortly after Boeing's bold advance came North American Aviation's 1964 report of more than one thousand pages detailing its own proposal for a "Manned Mars Landing and Return Mission Study." North American Aviation covered a broad range of topics, from a single or dual launch and rendezvous vehicles, to trajectories for Mars orbits, to diagrams of the landing capsule, and even included a secret fifth-volume supplement on "nuclear propulsion." The company was

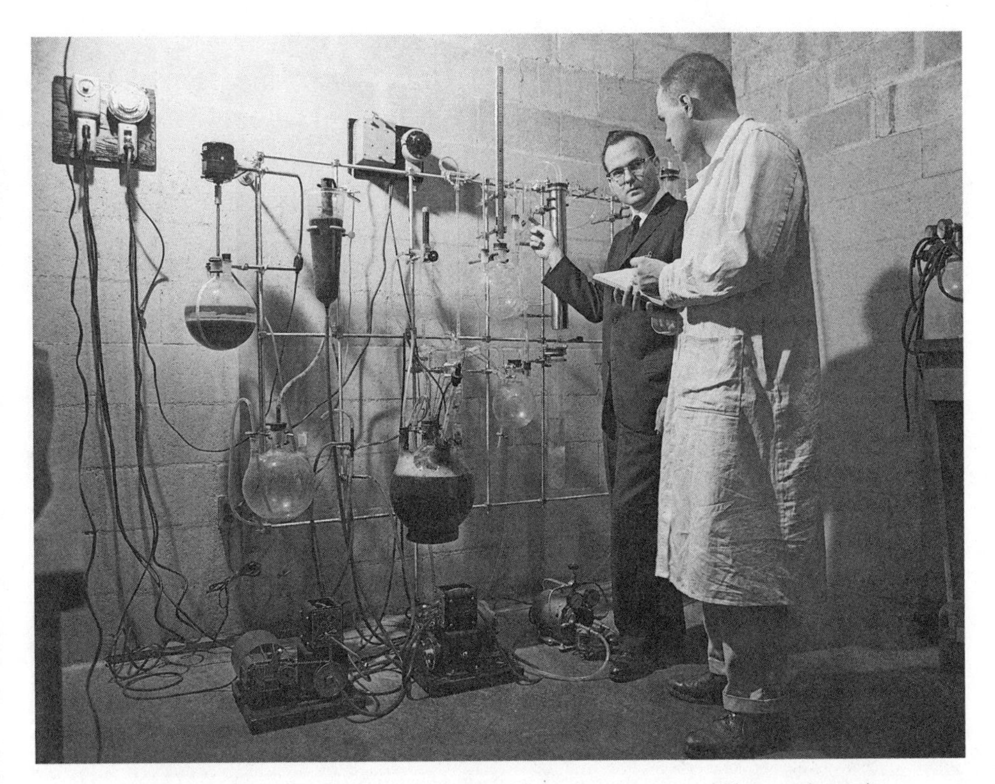

FIG. 1.8. Human waste recycling tests for Living in Space Experiments. nd BI219998 (p24543). Courtesy of Boeing Images.

clearly entering an early bid on the large enterprise of going to Mars: it dealt with various scenarios balancing the weight of food, air, and water versus crew size against several forms of chemical reprocessing. Its report admitted that its systems were only partially closed but argued that these were still appropriate for the twelve to eighteen months that a "there-and-back" interplanetary Mars mission would presumably take. A completely closed system was not included, the report said, "because of unsolved problems associated with recovery of human wastes." Obviously, the company's engineers had played with different biological options and were still undecided. "Both algae and higher plants can utilize the nitrogenous components of human wastes as nutrients," the report explained, but while algae grew faster, higher plants were "better gas exchangers, better human food sources, better food for supporting secondary animals, and less susceptible to disease than algae."[58] A year later still, General Dynamics's Convair Division agreed that the creation of a completely closed system was feasible in principle;

Fig. 1.9. Algae production for Boeing's living in space experiments. nd BI219976 (p23733). Courtesy of Boeing Images.

however, the "food-waste loop" remained the major technical hurdle due, they said, to incomplete studies of food's nutritional content, little knowledge about nutrients' journey through food chains, and basic ignorance about the content of biological wastes. In a moment of understated frustration, General Dynamics's engineers complained that such basic units of biology as "urine and feces are prime examples of poorly defined substances."[59]

Little dismayed by its earlier failure, Boeing continued its research into complete life-support systems based around a biological solution. Through the 1960s, Boeing worked to understand human waste and how it might be recycled through a closed biological system (see fig. 1.8). By 1967 it had developed an experimental system to process human waste with algae (see fig. 1.9). In fact, in an effort to demonstrate this system's impeccable functionality, the algae were then used to bake cupcakes—this, at least, suggested the company's pertinent public relations campaign (see fig. 1.10). There was certainly reason to expect

Fig. 1.10. Consuming algae cupcakes created for the Living in Space Experiments. nd BI219982 (p23969). Courtesy of Boeing Images.

that hiding the algae and their suspicious upbringing in cupcakes, which were completely unattainable in space, would be more promotionally effective than realistic images of a green puree of doubtful taste and odor.

THE DIFFICULTIES OF BIOLOGY AT NASA

Looking at these various endeavors within and without NASA, the life sciences emerge as an unexpected star of the early years of the American space program. The biologist "is of particular value in space," said Dale Jenkins, chairman of NASA's Office of Space Science and Applications, in 1966. Biologists and their knowledge were in high demand at NASA, not least because the problems of creating a life-support system for long-term space habitability remained elusive. While "considerable progress has been made on the engineering side of the program, there are still many unsolved microbiological problems," a NASA microbiologist, Judd Wilkins, explained to a conference in 1966.[60] NASA and its military-industrial contractors, as we have seen, knew that in addition to microbes, far more research was also needed on higher organismic levels: encompassing nearly every feature of life, Jenkins noted that the space age still required "basic studies in the low gravity environment [that] include research on single cells, plants, and animals, and include specialized studies of growth, aging, longevity, embryology, genetics, physiology, and behavior." Both the space effort and the life sciences were expected to benefit from such wide-ranging studies. Not only did NASA need this research, the new realm of space promised to reveal fundamentally new knowledge across many biological fields. An orbiting space laboratory, for example, offered a chance for a better understanding "of life processes and the role of the environment in maintaining normal organization and function of living systems." It would have to specialize in "experiments with environmental factors, such as weightlessness and decreased gravity, and removal from most Earth rotational periodicities affecting biorhythms." Of course, throughout the 1960s many of the projects came down to NASA's most immediate concern: "biological studies on algae and bacteria are needed in space to determine the effects of the space environment on the growth and effectiveness of these organisms in bioregenerative life support systems."[61]

Not only did biology shape the parameters of future missions, but future mission priorities also shaped what *biology* meant within NASA. NASA was an organization that, at its height, commanded about 5 percent of the scientific and engineering professionals in the United States. However, they overwhelmingly came from the various physical sciences, and to them biology was a strange science. The markedly different intellectual and scientific traditions of these disciplines became especially apparent once they were forced into joint projects, such as the creation of artificial environments. Charles Donlan, the associate director of Langley, allegedly said "that the language spoken by the biology people was difficult to understand in terms of specific objectives and possible rewards."[62] NASA's engineers were taken aback to learn how apparently little biologists knew about biological systems. As an early report to NASA noted, "Living organisms, whether man, amoeba or bread-mold, represent enormously complex, highly integrated systems whose nature is understood only fragmentarily and dimly by scientists." The epistemological foundations of the life sciences were difficult to grasp for outsiders, and the report explained, more than a little annoyed, how "there are few generalizations or laws in biology which have predictive value; hence, specific answers to new problems can usually be obtained only empirically." This was in stark contrast to subjects like engineering, "where known materials are combined according to exact laws." As the memorandum concluded, "Biology is not this way."[63]

Ecologists faced a similar situation in the 1950s when they cooperated with the Atomic Energy Commission to understand the movement of radioactivity through organisms and environments. As historian Stephen Bocking described, by the late 1960s, the Atomic Energy Commission hosted one of the largest groups of systems ecologists who benefited from stable patronage and the association with atomic energy but who also eschewed dealing with practical complexity in favor of the reductionism afforded by studying energy flow within isolated natural systems like ponds and constantly seeking to build grand theories in reflection of their nuclear physicists hosts. Ecology had to "be 'respectable' in the eyes of the physical scientists," one of the leading ecologists stressed.[64] Life scientists working within NASA's closed environments were equally troubled by how to relate that knowledge to their suspicious engineering colleagues. NASA's life scientists insisted on their specific competence and pointed out that "the variability of biological organisms and their complex processes make biological research difficult." This did not render their efforts superfluous but demonstrated, in contrast, "the need for a highly trained biologist to conduct experiments and observe results." These experiences readily fit into a larger history that sees biologists struggle to specify and then measure the phenomenon in question at the level of the molecule, cell, organisms, or ecosystem. In light of the remarkable success of their physicist colleagues in the twentieth century, biologists faced the challenge to defend the legitimacy of their own field's methodology, whether it was in the fields around nuclear reactors or inside life-support systems at NASA.[65] But, of course, the requirements of space flight played into the hands of the life scientists. The drive to take life into space meant that the life sciences were everywhere in the space age. In France, NASA reported, "Hector the Rat" starred as the test subject through a series of bio-astronautic experiments. The United States celebrated the studies and early flights of monkeys and primates. The Soviet Union learned much from their famed national model organism, dogs, who became national heroes so famous they starred on stamps.[66]

The importance of animal studies was undisputed, as Jenkins's report of 1966 emphasized: "Since the true consequences of prolonged space missions are not likely to be resolved through human monitoring and human experiments alone, the problem must be approached through experimental biology and animal studies." Facing the challenge of understanding complex biological actions and interactions at all scales from the microscopic up to the environment, the space biologist was expected to come armed with "broad training and experience in physiology, biochemistry, and cytology applicable to studies on plant and animal cell systems."[67]

The urgency of the space race, however, dictated where NASA's priorities must lie. Writing from Ames in October 1966, Billingham underlined "that basic and applied research on animals, particularly the larger species, is more important than cellular biology, experiments on micro-organisms, and exobiology, as far as the manned space station is concerned." The factual importance, however, did not immediately translate into recognition and acceptance within the laboratory. Reflecting back on the early years, associate administrator for Office of Space Science and Applications, John Naugle, recalled how life scientists felt that "their research objectives [were] usually not understood and often looked down upon by the physical scientists." It did not make things easier, Naugle thought, that the life sciences themselves were at odds with one another. Since each section of NASA enrolled its own life scientists, "there was also a continuing hastle between bioscientists in OSS [Office of Space Science], the aeronautical life scientists in OAST, and the 'flight surgeons'" in the Office of Manned Space Flight. It was probably small comfort to know that their Soviet rivals faced the same issue. Iosef I. Gitelson, one of the leading figures of the Soviet project to build an artificial environment, commented as late as 1992 that it was "common knowledge that the attitude of medical people and biologists differs from the attitude of the designers" when it came to incorporating controlled environmental systems into a space vehicle.[68]

FROM THE AQUA-HAMSTER TO PLUTONIUM WATER RECYCLERS

In the excitement of the 1960s, expertise, ideas, and interest flowed into NASA also in unsolicited ways, from sources far beyond the usual contractors.

Numerous people offered solutions to some part of the artificial environment dilemma. T. Wydeven and E. Smith, for example, built a contained water-vapor electrolysis cell to supply oxygen back to astronauts from the water "obtained from expired and perspired water vapor." Another idea was to combine two waste gases, methane and carbon dioxide, and convert them into vinegar.[69] *Popular Mechanics* reported on the work of General Electric chemical engineer Walter L. Robb, whose silicon membrane could be used as a gill to let mammals breathe air extracted from water. Robb made a splash when he built a mammalian environment inside a fish tank to create the "aqua-hamster." Finally, there were enterprising individuals like Dr. J. G. Schnitzer, who took the liberty of writing directly to Wernher von Braun on behalf of the Hygiene Work Circle of the Free Union of German Dentists. In this letter, which was translated by NASA into English, Schnitzer offered his services to develop "controlled decomposition" systems for long space flights or lunar bases. The persistent inquiries eventually ended up on the desk of Harold Klein, who wanted to know whether Schnitzer was a "responsible scientist" or just a crank. Klein queried whether Schnitzer's suggestion of "controlled decomposition" was just another name for the activated sludge sewage disposal system that NASA was already developing. Schnitzer replied that controlled decomposition was a vital part of any closed-loop system. It involved specifying the minimum loop required to recycle the waste products of "vegetable foods" and "fecalia" into a "fertile and healthy humus" to be used as fertilizer on farmland. Schnitzer claimed to have most of the data already to develop such systems. The only real major question remaining, Schnitzer said, was "the problem of miniaturization." To him, it was an open question, "*how small* such a bio-cycle can be constructed to still be able to function"—and that question plagued NASA as well.[70]

In short, by 1965 an array of practical proposals and pure flights of fancy floated through NASA's Life Science Directorate at Ames. Optimistic senior NASA administrators pushed ahead with the process of rationalizing what they saw as the next phase of the American adventure in space. Deputy NASA administrator Robert Seamans advocated an intensive sixty-day study to evaluate the need, requirements, and constraints of a permanently manned orbiting space station. As a result of this study, Ames declared itself ready by August 1966 to "provide a fully tested and reliable bioregenerative life-support system capable of supporting at least 10 men for 5 years" within a decade. Thus, in mid-September 1966 a new Space Station Steering Committee met with the Configuration Description Group from the Advanced Manned Mission Office of the Office of Manned Space Flight at NASA headquarters. The group agreed that a space station was a crucial precondition to any potential interplanetary

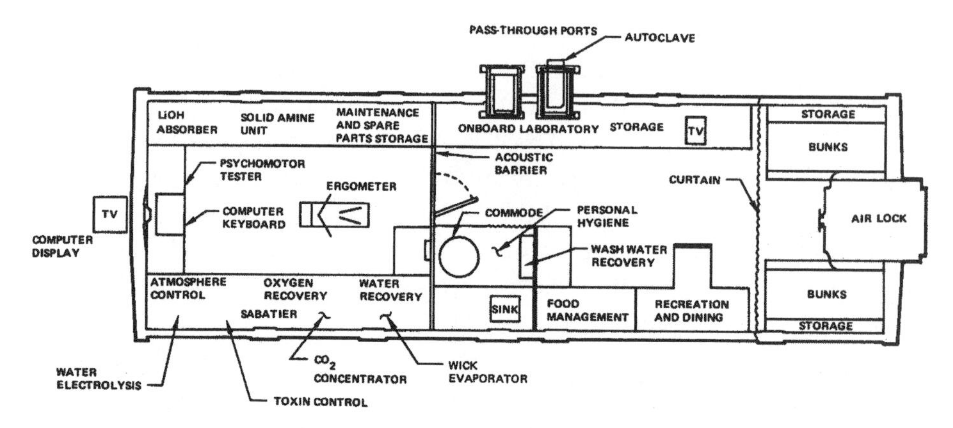

Fig. 1.11. "Space Station Simulator Arrangement for 90-Day Test." From J. K. Jackson, "Life Support Systems," in *Preliminary Results of an Operational 90-Day Manned Test of a Regenerative Life Support System*, NASA SP-261 (NASA: Langley Research Center, 1970), 40.

Fig. 1.12. "Waste Management System (Commode)." From Courtney A. Metzger, "Design and Development of the Vacuum-Distillation, Vapor-Filtered, Isotopic-Fueled Water Recovery System for the 90-Day Manned Simulator Test," in Albin O. Pearson and David C. Grana, *Preliminary Results of an Operational 90-Day Manned Test of a Regenerative Life Support System*, NASA SP-261 (NASA: Langley Research Center, 1971), 248.

flight, as it would serve to test-run all processes and provide a "demonstration of both the individual and complete system reliability."[71] Promoted to acting chief of the Biotechnology Division, John Billingham excitedly reported to his fellow leaders in the Life Sciences Division that the presentations of the Steering Committee had convinced their colleagues from the Mission Office. The NASA project would be a "22-foot, nine-man space station, orbiting at 55 degrees and an altitude of 200 nautical miles" that would "immediately" meet "nearly all the requirements of the various panels" NASA had consulted. It would host basic and applied biomedical and behavioral studies, test experimental life-support systems and environmental controls, and evaluate the reliability of protective systems for a crew. The key question that prompted a "vigorous discussion throughout" was whether the space station should rotate: while some astronomers wanted a stable non-rotating viewing platform, some meteorologists believed that a more habitable station equipped with artificial gravity would make maintenance easier and the scientists onboard more productive.[72]

Unfortunately for all protagonists involved, over the next years the hope that the United States would launch a permanent space station in the immediate future evaporated. The United States chose to escalate the conflict in Vietnam, so NASA's budget was slashed. The goal of a rotating or non-rotating American space station faded until 1983. The severe budget cuts of 1967 and 1968 hit the life sciences' projects extremely hard. The first hints came as early as September 1966: in between the first and second drafts of the major planning document for the manned space station, someone changed "all EC-LS [Environmental Control and Life-Support] System will be tested . . . in Apollo" to "all EC-LS Systems may be tested." The development of long-term life support was no longer a near certainty as it once was, but now only a hopeful speculation. By 1970 all that was left coagulated into the *Skylab* project that, from the start, was underfunded, underdeveloped, and yet constantly trimmed.[73]

But big science, multi-agency projects take almost as long to shut down as they do to start up. Already planned and built, McDonnell Douglas's Astronautics Division conducted a full-scale ninety-day "space station simulator" of its "regenerative life support system" in 1971 (see fig. 1.11). Four male volunteers spent the time in a sealed 4,100-cubic-feet chamber with all their food presupplied but their air and water recycled. It was a major project to prove the system's reliability. Once again, waste collection, management, and testing formed a major part of the system, with a newly designed commode (see fig. 1.12). A fecal sampler extracted small amounts of solid waste from the commode for medical testing, a no doubt confronting experience for the simulator participants. McDonnell Douglas particularly stressed the "monitoring" capability of the system that was

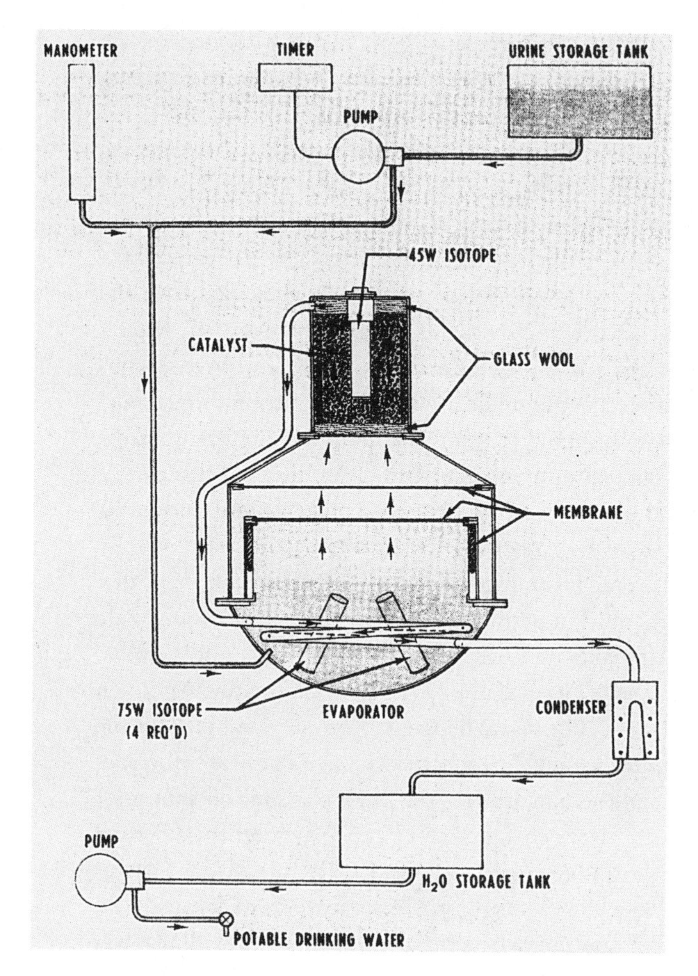

FIG. 1.13. "VD-VF Water Recovery System with Radioisotopes as Energy Source." Courtney A. Metzger, "Design and Operation of a Waste Management System for Fecal Collection and Sampling during the 90-Day Manned Simulator Test," In Albin O. Pearson and David C. Grana, *Preliminary Results of an Operational 90-Day Manned Test of a Regenerative Life Support System*, NASA SP-261 (NASA: Langley Research Center, 1971), 96.

able to detect and follow organic components, but also inorganic substances, microbiological organisms, and trace elements. Electrical power, however, was supplied externally, to run the hot and cold water supply and the electrolysis apparatus to process carbon dioxide. McDonnell Douglas especially celebrated their testing of a nuclear power source using "microspheres" of plutonium oxide ($^{238}PuO_2$), to run a water recovery subsystem (see fig. 1.13). This, of course, made radiation health an alarming issue, as the team admitted.[74] But the solution of this and other questions had to be postponed because just when an artificial environment had more or less successfully been tested, the United States lost interest, and much of the fire went out of the space age.

2

THE ALGATRON VERSUS THE FECAL BAG

A kind of "algae space race" developed.

—William Oswald, 1982

IN SPACE, NO ONE CAN HEAR YOU SCREAM—OR, FOR THAT MATTER, SHIT. In more than five hundred pages of retrospective account that Neil Armstrong, Edwin Aldrin, and Michael Collins provided of their 220,000-mile journey toward the moon and back, there is not one reported incident of anyone going to the bathroom. The popularly published Apollo 9 flight plan listed every operational maneuver of the first test of the lunar module, including the regular sleeping and eating periods, but made no mention of a poop period.[1] At most, some official accounts of the space program sanitize the biological reality by converting such moments into medicalized parlance. NASA's monumental history *Chariots of Apollo*, for example, noted that on Apollo 7 the crew found the "waste management system for collecting solid body wastes," as it is circumscribed, "adequate, though annoying," and specified that the crew had "had a total of only 12 defecations over a period of nearly 11 days." However, nowhere in the reports of Apollo 8, 9, 10, or 11 does one find such information. Some more popular accounts of the space age almost blush when they venture into the area of space sanitation: cofounder of the Australian Mars Society Guy Murphy told his teenage readers that it is still a "delicate issue of what an astronaut does if they need to visit the bathroom."[2] Likewise (though relegated to a footnote), the famous science writer and novelist Arthur C. Clarke rhetorically asked "the invariable question 'How *do* they manage?'" to which he answered, "Not very well."[3]

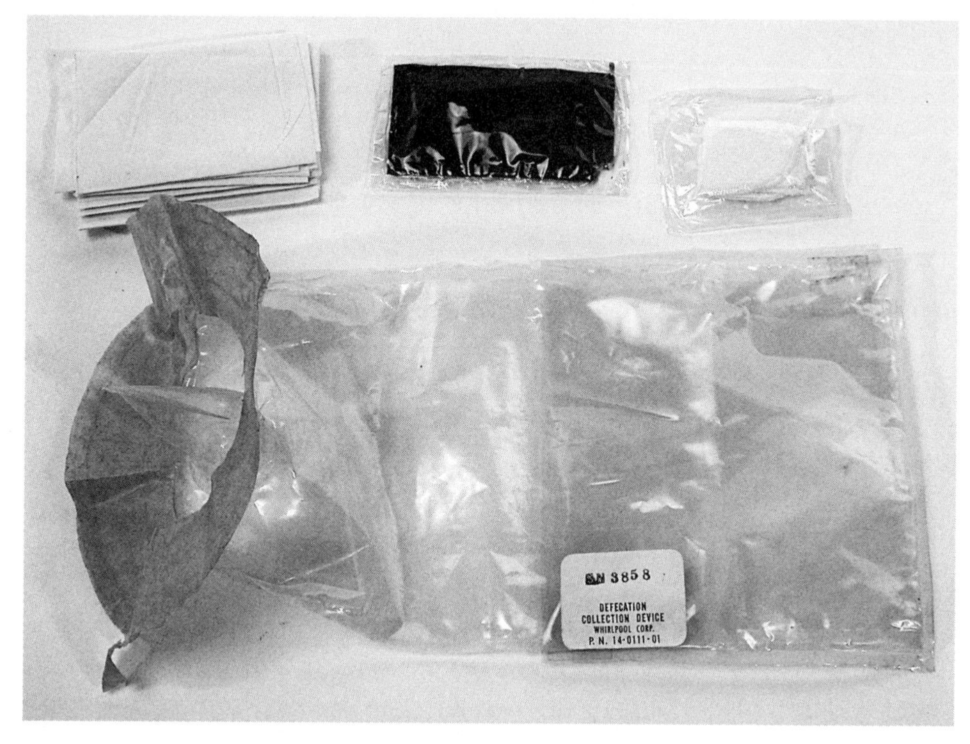

Fig. 2.1. Whirlpool Corp., "Apollo Fecal Collection Bag," c. 1969. Image courtesy of the National Air and Space Museum. Inventory no. A19750739000.

In contrast to discomforted adult audiences, the question of going to the bathroom in space is well known to be "the most popular question [astronauts get] from younger children," said astronaut Tim Peake in his 2017 memoir.[4] Indeed, some astronauts have offered rather forthright descriptions of the confronting and complex task, notably from Russell "Rusty" Schweickart of Apollo 9. Speaking at the end of a day of interviews in 1976 with Peter Warshall of the *Whole Earth Catalog*, Schweickart went where few had gone before to describe the operation of the fecal bag (see fig. 2.1):

> In Apollo, for feces you just stuck a plastic bag on your butt which was 6 inches in diameter—something like that—maybe a little bit less than 12 inches or so long and the mouth of it had a flange at the top with an adhesive on it, and you'd peel the coating off the adhesive and literally stick it to your butt. Hopefully centrally located. And if you think you know where your rear end is, you really find out, because you'd paste it on very

> carefully! So, you stick that to your butt, and then you go ahead and take a crap. But then the problem comes, because there's no particular reason whatsoever for the feces to separate from your rear end. So as a result the problem is left as an exercise to the student to peel the bag off and make sure everything stays within the bag, and get all wiped off. It's basically a one hour procedure.[5]

The procedure, in other words, was as long and involved as many other parts of an astronaut's daily rituals; it was complicated enough to require its own checklist.[6] Yet the daily evacuation is rarely given the same discussion as any other daily event. When it is discussed, the process is usually framed as moments of unflappable masculinity in the face of adversity. The bathroom suddenly turns into an obstacle to be overcome. Take the well-known story of how in 1961 astronaut Alan Shepard, who became the first American in space, soiled himself as he waited firmly fastened in his seat for launch, which had been delayed for several hours. Novelist Tom Wolfe turned this moment of social embarrassment into an heroic feat: "Imperturbable at every juncture," Wolfe exclaimed.[7] Fifty years later, after he returned from his yearlong space flight in 2015, astronaut Scott Kelly recalled chuckling to himself as he got "to diaper up" at the thought that "I wouldn't have to be in diapers again until much later in life." More official accounts, like that of NASA historians David Compton and Charles D. Benson, euphemistically summarize the culture of the astronauts by remarking how "quite a lot of minor inconvenience could be tolerated by a man on his way to the moon." Schweickart's tale belies any claim that an astronaut's toilet was merely minor inconvenience, and so Compton and Benson's euphemism must be understood as a conscious attempt by NASA to sanitize its own history. More realistically, the popular writer Andrew Chaikin observed that the practicality of going to the bathroom ruined the image of the heroic astronaut. It was certainly "not the image of the Intrepid Lunar Explorer" anyone wanted, Chaikin noted as he described how Ken Mattingly on the 1972 Apollo 16 flight was "slurping down a plastic bag of juice while hooked up to the urine collection bag, with a fecal collection bag flypapered to his rear end." In the same vein, Mike Mullane, famous for his space shuttle flights during the 1980s, admitted that having to be toilet trained and fitted for a condom catheter "took a lot of the glamor out of being an astronaut."[8] Censoring bodily functions in standard histories has thus preserved the glamor of space—and barred a host of uncomfortable memories.

As good a summary as any, in the final scene of the 2015 movie *The Martian*, astronaut Mark Watney visits a class of future space cadets and shares with them how the other crew members accidentally had left him on Mars and how he survived by "farming in his own shit." He assured his audience that it actually was "worse than it sounds" and concluded, "so let's not ever talk about that ever again."[9] Yet that's exactly what this chapter is going to do because not only was the actual act of going to the bathroom a critical problem in the early space age but so too was the creation of a whole sanitation system. Waste management was every bit as important as rocket launches and moon walks.

DIRT AND DISORDER IN SPACE AND ON EARTH

Sanitary engineering has been a significant part of human history. Ideas about, technologies for, and the social relations of sanitation exist in a close and reciprocal relationship with a society's ideas of class, gender, and social order.[10] Whether it is trash or excrement, the outcast or the shunned, anthropologists and sociologists have argued that what people keep and what they discard is indicative of what is considered valuable and what is not. Changing conceptions of waste and how it should be dealt with are thus always situated in historical contexts. "Dirt is essentially disorder," Mary Douglas observed, and "eliminating it is not a negative moment, but a positive effort to organize the environment."[11]

Human waste, in the layered meanings of the expletive *shit*, philosopher Dominique Laporte argued, is a broad category ranging from the physical through the semiotic and psychological.[12] What is considered waste and what people decide to do with it has shaped urban and social environments over the last century. According to Laporte, a critical moment in history was when a French royal edict announced in 1539, "to each his feces." From that point on, households no longer threw their waste into the public gutter but were obliged to store it, whether it was "refuse, offals, or putrefactions, as well as waters whatever their nature." As historian Susan Hanley noted, in premodern Japan human excreta used for fertilization had a substantial value: a year's worth of urine and feces equaled a month's wages.[13] Nightsoil men collected and transported the materials out to farmland, where it was reintegrated into the society's ecology and economy. "Human fertilizer is without equal," one mid-nineteenth-century Frenchman enthusiastically declared—sewage was successfully used to treat farmland just outside Paris for over a century, until the 1980s.[14] In twentieth-century America, however, excrement became considered waste that had to be removed alongside other types of garbage, thereby equating the broken chair with the organic stool. Human excrement became one of the most repellant substances, a potent inverted symbol of value. The artist Pietro Manzoni famously shocked the world in 1962 by distributing cans of his own shit as a conceptual protest against the commodification of art and the artist; the price tag he put on each 30-gram can was 30 grams of gold. Nearly forty years later, Wim Delvoye's installation *Cloaca* in the Museum van Hedendaagse Kunst in Antwerp was fed a meal from a plate and then processed the food completely, including excreting the waste from the other end. These daring and confronting artists saw human waste as the ultimate anti-value substance, commenting on the nature of twentieth-century society and the artificial valuations of artistic products.

But in the space age, waste looked very different. The goal of developing sanitary systems as part of the wider effort in life support were high on NASA's priority list. Under extreme time and weight pressure, the principal choice of NASA was between "collection and storage" and "bioregeneration" of waste. This chapter elaborates on that choice by looking specifically at the example of the fecal bag versus the Algatron, an algae-based waste recycling system developed in Berkeley by a pair of sanitary engineers at the Sanitary Engineering Department of the University of California, Berkeley.

THE FECAL BAG

In chapter 1, we saw how life scientists teamed up with engineers around 1960 to develop life-support systems for future NASA projects, namely a space station and a mission to Mars. However, amid the deteriorating social and political landscape of the United States, President Lyndon Johnson's administration refused to entertain any post-Apollo plans and all visions of longer flights were canceled. Instead, the decision was made to go to the moon and back from a terrestrial launch, which shortened the time in space to less than ten days.[15] This dramatically altered the course of NASA and its industrial contractors. Bioregenerative environments became superfluous, and biomedical experiments concerning the long-term physiological effects of living in space were discontinued "on the grounds that they were not critical to Apollo." For the Gemini program, veterinarians were enlisted to concoct a highly digestible astronaut diet to reduce the volume of excrement, and adult diapers were used to deal with the inevitable, if much reduced, feces. The practice by Apollo 9's 1969 flight (Rusty Schweickart's mission) was that urine was dumped "overboard" while the "plastic defecation bags . . . [were] sealed after use and stowed in empty food containers for post-flight analysis."[16]

In short, Americans went to the moon equipped with nothing more than a sealable plastic container, better known as a fecal bag. It seems a remarkable case of technological devolution. For some time now, historians of technology have found it illuminating to explain why some technologies are adopted while others are abandoned—stating that technologies are adopted because they're somehow better than their rivals begs the interesting question of what in each case was considered better and why. Ruth Swartz Cowan's work on the gas-powered refrigerator losing to its electric rival and Rudi Volti and David Kirsch's study of the electric automobile losing to gas automobiles provide instructive examples of how technological decisions are always negotiated in social contexts.[17] Many factors likely contributed to the fact that the fecal bag prevailed. On the surface, the choice was driven by pragmatic consideration: bioregenerative systems were considered essential for longer-term space missions, whereas on short flights the additional weight of the technology was not worth the trouble. In addition, when the United States said it would land on the moon within the decade, such systems were not yet working properly and, therefore, given the self-made time pressure, were dropped from the list of priorities. However, there was a second layer: in view of the 1960s obsession with hygiene, cleanliness, and disinfection, it may have been difficult to favor a decision for biological systems that acknowledged human excrement as part of ecological cycles and, eventually, as the raw material for sustenance. Whichever was more important, collect-and-storage systems prevailed.

The fecal bag became viewed in retrospect as the natural method of dealing with feces in space; however, for many scientists and engineers within NASA, and also for the astronauts themselves, the fecal bag was really a stop-gap choice that nobody was particularly happy with, not least since the collection bags did not work very well. In fact, they might be the epitome of Donald MacKenzie's snappy phrase that "technologies . . . may be best because they have triumphed, rather than triumphed because they are best."[18] Using a bag for "defecation in space is an art," Jim Irwin of Apollo 15 said with some justification. Beyond the troublesome procedure of using the bag, its storage function was not assured: astronaut William Pogue reported that when feces inadvertently escaped the bag, it caused many a "ribald comment" among the crew. Science popularizer Mary Roach reports several outrageous incidents of this kind, including the Apollo 17 crew laughing about a "turd" floating around the command module. Frank Borman on Apollo 8 "ended up urinating all over the spacecraft" because, he later said, "as usual the UCD [urine collection device] did not work properly." As histories of zippers and condoms testify, when technologies meet genitalia they seem destined to break or malfunction at the most inconvenient moments.[19] Even when they worked, the smell, feel, look, process, and psychology of using fecal bags was so odious that some early astronauts preferred starvation to eating and subsequently having to use the bags. In December 1965, on Gemini 7, Frank Borman managed to go nine days without defecating, a new record that surpassed the earlier 119-hour flight of cosmonaut Valery Bykovsky in June 1963.[20]

One can hardly blame them: once the excrement was inside the fecal bags, they had to be churned by hand with an antibacterial agent before storage.[21] Even now the much-improved mechanisms are supplemented by tricks learned by the astronauts themselves: just before he spent 2015 on board the International Space Station, Scott Kelly asked his Russian hosts why there was dill in everything they ate. "Dill gets rid of farts," he was told—a worthwhile goal, he noted sagely, "before being sealed into a small tin can."[22]

THE ALGATRON

One of the most evocative alternatives to dealing with human biological waste in space was called the Algatron. The Algatron was an ecological system of algae cultures that provided "conversion of liquid and solid wastes into purified water and other useful end-products, including food," and was planned to permit "the establishment of a complete ecological system in lunar bases, in manned deep space probes, and in permanently orbited earth stations." The idea was to provide "humans sealed within an isolated capsule" with all they needed for "indefinitely long periods of time."[23]

In the late 1950s a pair of sanitary engineers at Berkeley began to develop the system as part of the flurry of creativity and innovation indicative of the early space program. The development was funded, up to the mid-1960s, through research grants from the Air Force Cambridge Research Laboratories and Berkeley's College of Engineering and School of Public Health. Thus, the Algatron was one of those devices developed by external partners to meet the demands of the American space program. It was an early instantiation of exactly the kind of self-sustained ecosystems using closed cycles of matter and energy that NASA was looking for at the time.[24]

The project came from a collaboration between sanitation engineer William J. Oswald and biologist Clarence G. Golueke. A young officer stationed in England during World War II, Oswald was in charge of water safety at one of the Normandy beaches. After graduating with a degree in civil engineering from Berkeley in 1951 on the G.I. Bill, Oswald specialized in sanitation engineering, public health, and biology. He joined the Berkeley faculty after finishing his PhD in 1957, during which he had experimented with the treatment of wastewater using algal systems.[25] Golueke, in contrast, was an expert in the science of composting and other techniques of solid-waste management. Oswald and Golueke's earlier work in water treatment facilities used large open algae ponds. From this perspective, their work on regenerative systems for space was mostly an exercise in miniaturization. Algal research formed a major portion of the work at the

Fig. 2.2. William Oswald with Algatron, Richmond Field Station, April 21, 1965. UARC PIC 27A—NEG 0858M (20A-21). Bancroft Library, University of California, Berkeley. Reproduced with permission.

Sanitary Engineering Research Laboratory at Berkeley in those years. Oswald and Golueke had looked at desalinization with algae, an algal treatment of agricultural and organic industrial waste, and the "nutritive value and disease-transmitting potential of sewage-grown algae as an animal feed supplement." When they tried to provide proof of their concept, they readily reported, already in 1964, that they had successfully maintained "continuous cultures for more than five years," putting to rest concerns that bacteria would eventually take over.[26]

To Oswald and Golueke, it seemed obvious that the space program needed a similar waste recovery system in miniature. They specifically chided NASA for adopting "such startling 'do it yourself' processes as . . . the handling of feces with a specially constructed glove"—this was the precursor to the fecal bag. However, they admitted that few sanitary engineers had "taken an interest in closed system problems." The challenge to miniaturize and enclose a sanitation system was no small order. Like so much about humanity's initial exploration beyond Earth, almost all the usual assumptions about how an environment functioned did not apply to a sanitation system for space. Not least, it had to accommodate a full set of environmental parameters: gravity (at least "1/3G"), pressure (at least "1/2atm"), temperature ("22°C"), oxygen (35–40 percent), relative humidity (50–60 percent), noise level ("10db"), air movement ("20fpm"), light ("0–50ft-candles"), and odors ("none of reduced substances"). Within those boundaries, they explained, the human organism thrived, consumed energy, and produced all manner of waste. In addition to solid feces, these included liquids, such as "urine, perspiration, skin oils, nasal mucous, sputum, saliva, tears, and semen emissions," and "gaseous wastes," such as "carbon dioxide, water vapor, regurgitated gas, and flatus." Oswald still maintained as late as 1968 that they needed to find ways to return the nutrients lost via these wastes to the human via a "mammalian-algal-bacterial closed ecological system."[27]

Between 1957 and 1962, Oswald and Golueke built a scale model of the eventual component system, then tested and enlarged their design (see fig. 2.2). In the same period, they started to speak of their system as the Algatron, a name that consciously emulated the famous Berkeley cyclotron. They were by far not the only ones to explicitly draw this connection. In fact, the 1950s and 1960s saw the emergence of an entire family of *-trons* in biology: starting with the phytotron and followed by the Climatron, Biotron, and Ecotron, these *-trons* were all increasingly sophisticated facilities of comprehensively controlled environment research.[28] They were mostly affiliated with plant physiology departments and rather generously funded. Both the projects and the facilities are mostly forgotten today, but in the 1960s the advent of new technological tools like computers and controlled environments were considered as important as "new concepts in biochemistry and genetics," as Harvard plant scientist Kenneth V. Thimann declared at a meeting of the Panel for the Planet Sciences at the National Academy of Sciences in 1966.[29]

All of these facilities shared an approach to biology as a cybernetic science with the technologies of control integral to the study of the processes of life—and even life itself was interpreted as an integrated amalgam of systems. There were also a number of smaller associated devices that took up this naming and research philosophy, like the assimitron, which measured the carbon-dioxide uptake of a canopy; the dasotron, which studied small ecologies; and the rhizotron, which is a chamber for viewing tree roots and various arthropods that live underground; and even the Eggatron, which computerized the time of hens' laying and the size of the eggs.[30] In 1959, it seemed to Donald Griffin, the discoverer of echolocation, that only a cycletron and a marinetron for water biology remained to be built to complete biologists' experimental control over the natural world. By the 1970s, however, the *-trons* of biology largely disappeared as part of the general fading of whole-organism biological sciences, even as whole systems, cybernetics, and alternative technologies gained favor with countercultural environmentalism.[31]

The Algatron was a special instance of the *-tron* family, which until around 1966 grew in technological and biological complexity. It was a controlled environment within a controlled environment, a closed-loop algal system within the larger systems of life maintenance of a space habitat. At the heart of the system were continuously growing cultures of unicellular green algae, one of the favorite model organisms in plant sciences at the time.[32] Oswald and Golueke ended up working with Mitchell Sabanas of the Lawrence Radiation Laboratory at Berkeley, an important collaboration. Back in 1962, Sabanas had tested life-support systems by enclosing an unrestrained primate for twenty-five days inside a sealed chamber of about a third of a cubic meter, with unlimited food and water alongside a mechanical carbon dioxide and sanitation processing system.[33] Oswald subsequently built a scale model of the actual Algatron in 1963 and enclosed a mouse in this system for six weeks to observe the longer-term viability of the device. In another run that year, a "closed algae-aerobic bacterial-anaerobic system was recycled for a period of 250 days without adversely affecting the cultures."[34] At the same time, a variety of experiments with mice, monkeys, and an array of algae species worked to specify the necessary parameters to create a stable gas-exchange system, including the rates of growth, photosynthesis, and respiration, as well as the changing protein, lipid, and carbohydrate composition of algae over several hundred generations.[35] Still, no part of the extant plans and models of the Algatron provides information of how the defecation itself would proceed; rather, the system offered through its recycling process an alternative

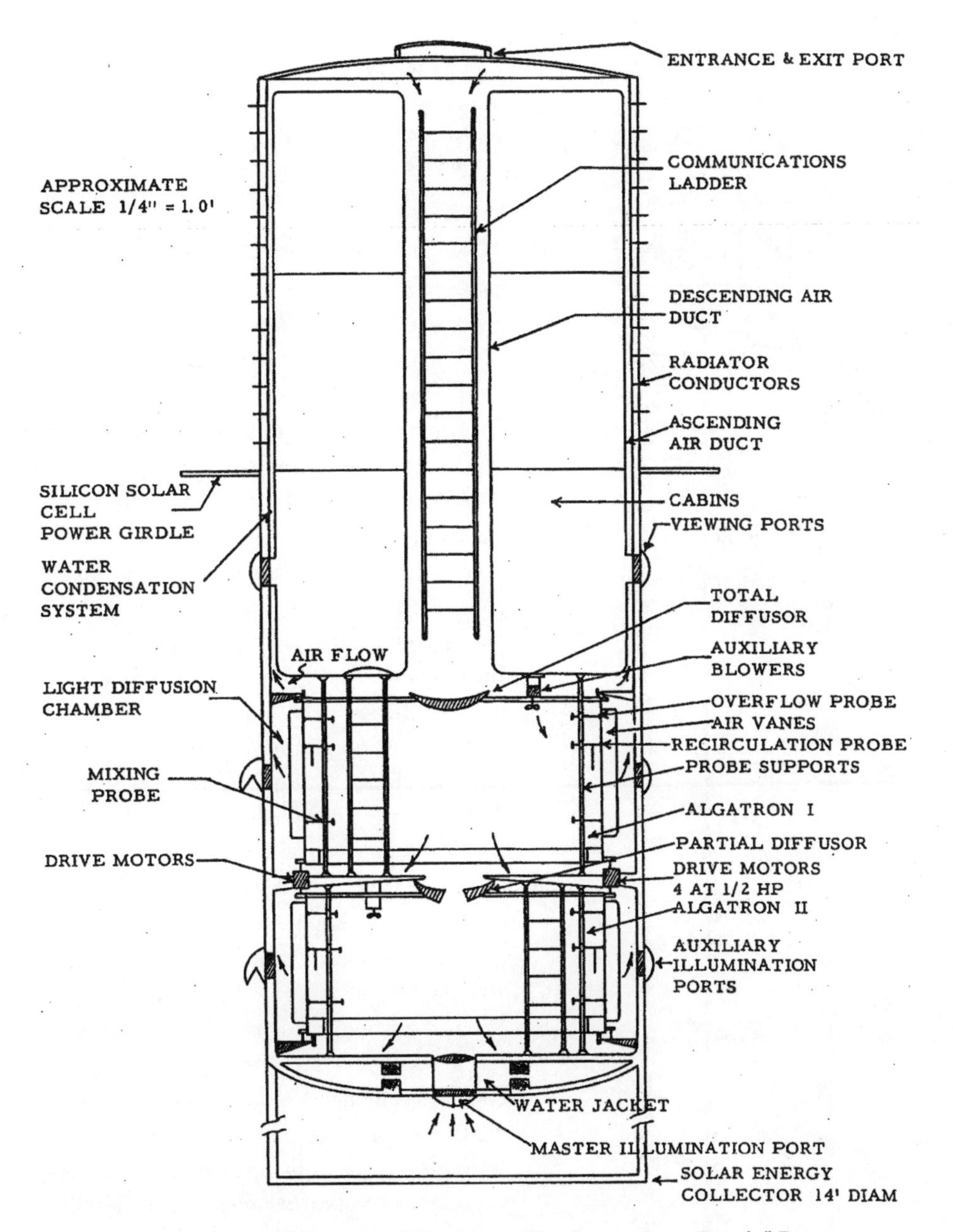

FIG. 2.3. "Incorporation of Algatron and Illumination Chamber in a Space Capsule." From William J. Oswald, Clarence G. Golueke, and Donald O. Horning, "Closed Ecological Systems," *Journal of the Sanitary Engineering Division, ASCE* SA4 (August 1965): 43. With permission from ASCE.

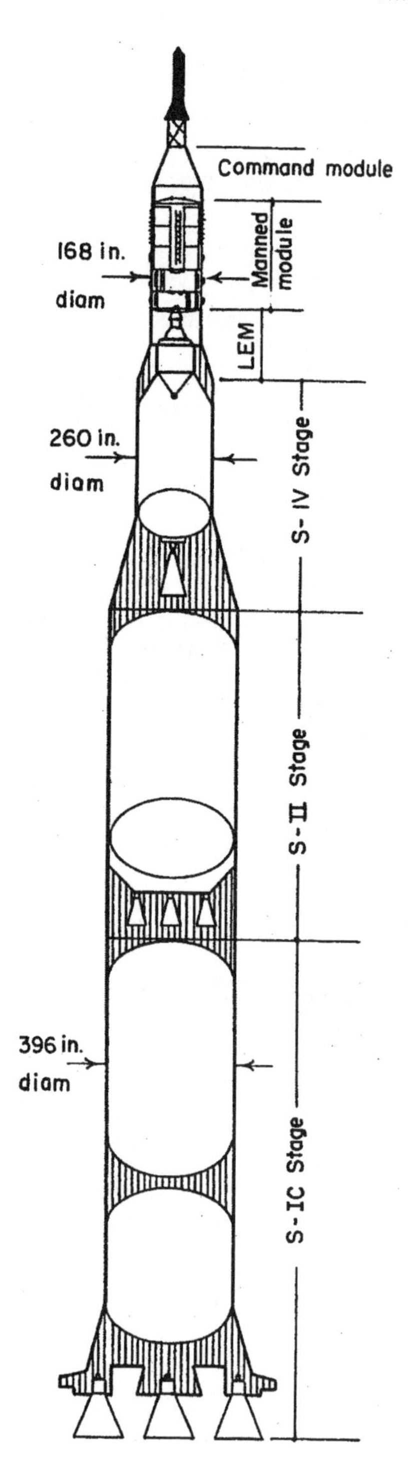

Fig. 2.4. "Closed Ecological System Mounted in a C-5 Rocket." From William J. Oswald, Clarence G. Golueke, and Donald O. Horning, "Closed Ecological Systems," *Journal of the Sanitary Engineering Division, ASCE* SA4 (August 1965): 44. With permission from ASCE.

to storing the product in an empty food compartment and taking it back home.

The final form of the Algatron that arrived around 1966 is worth a more detailed description to better appreciate its complexity and the engineering and biological challenges involved. It would sit in between the command module and the LEM (lunar excursion module) (see figs. 2.3 and 2.4) and comprised a stacked pair of double-walled transparent cylinders occupying most of the diameter of the proposed crewed module (168 inches). They were exact copies of each other but spun in opposite directions to avoid placing a torque on the spacecraft. The rotation was driven by an electric motor, which was powered by external "silicon solar cells" arranged in a "girdle" around the command module. Light ports allowed sunlight onto the cylinders, while nutrients (human waste) were introduced via the "overflow or decanting scoop" made of Teflon, which sat just inside the surface of the cylinder. This is also where the algae cultures would grow: on the interior surfaces. The culture was mixed by a static probe as the cylinder rotated. To aid circulation of carbon dioxide and control the temperature of the thin film of algae culture (not all the radiant solar energy would be converted to algae growth), a small gap between the command module and the rotating cylinder allowed for the "rapid movement of dry air."[36] It was a sophisticated design, and although it never made its way into space, the principal approach was picked up by many successor models—in particular the decision to work the system with cultures of green algae.

ALGAE AS THE WORKING MACHINERY OF LIFE SUPPORT

The use of green algae for bioregenerative systems was no coincidence. In the first half of the twentieth century, algae had led a quiet existence as the main model organism of photosynthesis research. They became, however, an almost utopian substance around the 1950s, promising to cure world hunger, solve its energy problems, and convert efficiently the global glut of sewage. In principle, the Algatron was a miniaturized sanitation system for crewed missions in space to solve for astronauts what Oswald's own high-rate Oswald Ponds have solved for increasing urban and suburban wastewater treatment. In both, algae was the miracle: Oswald's biographer John Benemann characterized Oswald's position as always, "when in trouble, algae to the rescue."[37]

Small, entirely unspectacular freshwater algae cells of the genus *Chlorella* came to enjoy particular fame. It had been introduced as a model organism for photosynthesis research in 1919 by German cell physiologist Otto H. Warburg and soon became the favorite research object for this field up to the 1960s. Chlorella was initially considered a reasonable choice because it was believed to be the smallest organism capable of the full photosynthetic reaction, but chlorella also had practical benefits. It grew in large quantities on light, minerals, and carbon dioxide, and because the alga's chloroplast occupied half the cell volume, its photosynthetic yields were relatively high. In the course of time, however, chlorella demonstrated an unsuspected metabolic plasticity, which forced physicochemical-oriented physiologists to acknowledge that there were multiple paths of photosynthesis that were flexibly activated depending on environmental circumstances.[38]

One of the discoveries during these investigations in the 1940s was that the main product of chlorella's photosynthesis under natural circumstances was not carbohydrates but protein. At a time of rising Neo-Malthusian anxieties in the United States, in which scenarios of overpopulation and hunger abounded, these findings offered hope. The American wartime campaign for the four freedoms, including freedom from want, was extended to become an integral part of the emerging Cold War, with fears of starving nations falling to communism. In that ideological context, chlorella appeared a potentially powerful weapon—if only it could be grown in sufficient quantities.[39] According to its boosters, chlorella was able to provide valuable food protein grown on a fraction of the arable land required for beef or grain.[40] This was the beginning of decade-long attempts to mass-culture chlorella as a nutritional supplement.

Important insights into chlorella came from the foundational work of the plant physiologist Jack Myers at the University of Texas in the 1950s. Myers graduated from the University of Minnesota in 1939 and taught at the University of Texas from 1941 until 1980. Widely known as an expert of photosynthesis and the physiology of algae, Myers was among the first scientists to seriously propose bioregenerative life-support systems for space based on using algae to convert carbon dioxide.[41] The controlled, continuous growth of large, long-term algae cultures in suspension seemed entirely possible once the issue of gas exchange through the liquid was technically solved. The chief difficulty, Myers explained, was "to miniaturize and reconstruct in small volume some approach to the balance of the biological world." By the mid-1950s, mass-culture plants of chlorella were established in countries all over the world, including—besides the United States—East Germany, Hungary, Israel, Japan, Mexico, Poland, Czechoslovakia, and the Soviet Union.[42] Historian Warren Belasco describes a world that sought to maximize the incident solar energy and economically deliver calories and nutrients via standardized foods to an expanding population. Yet, in the end, most people in Europe and America proved unable to develop a taste for the algae; and with the exception of East Asian countries, the mass-cultured algae were mostly used for feeding fish.[43] American plant physiologist Charles Stacy French was certainly an exception when he declared in a popular 1962 essay that chlorella powder "was extremely good when mixed into ice cream or added to the flour used in making bread." He found, in contrast to most of his colleagues, that it had a "rather pleasant vegetable taste with a touch of the odor of violets."[44]

The same Cold War era of technological optimism that held the power of algae to transform hunger into abundance equally considered the Algatron to be an elegant solution to the problem of how to return solid, liquid, and gaseous wastes in space to biological circulation. The hope was that algae could forge a link between the intake and output cycles of multiple organisms within an ecosystem, even in space. Oswald and Golueke were optimistic that this was going to work: "The bond between man and his physiological wastes is so unbreakable," they maintained, that even leaving Earth for space could not "interrupt the cycle of life."[45] Oswald and Golueke stressed how their bioregenerative system would reduce the space vehicle's launching weight. In fact, "only that amount needed to meet the demands of the crew for a day or two" would have to be taken aboard before the algal culture would return water, food, and oxygen for further consumption. They even offered estimated values of oxygen, food, and water requirement for periods from one day to ten thousand days.[46]

Oswald and Golueke were remarkably clear about how they saw their system. They maintained in 1965 that the Algatron was "a miniature version of the grand scale terrestrial ecological system of which we are a part"; the basic principles of the two systems were the same, they explained, differing "only [in] the size and variety of their constituents." This vision of creating small closed versions of terrestrial ecosystems to enable people to live in space formed much of NASA's vision throughout the 1960s, as the welcome pamphlet for the Conference on the Closed Life Support System at the Ames Research Center in April 1966 displayed (see fig. 2.5). Out of necessity, a reduced ecological system would be brought into space, but within these limits one still should try and keep the system as variable as possible. Ecological research had shown, Oswald and Golueke pointed out, that "a system becomes more stable as the number and varieties of individual organisms increase."[47] Life-support systems, hence, fared better if they included multiple organisms from different species alongside the human mission crew.

The creators of the Algatron also questioned the central role of humans in their system.[48] It became a widespread attitude that perhaps their most important function on board a spacecraft was to ensure the system's reliability and robustness—humans were needed to identify and fix the problems in situ that were bound to arise, serving more as repairmen than innovators. Oswald and Golueke went even further: they were not too optimistic about the crew being up to the challenge. "Man is biologically the least reliable component of the [spaceship] system," Oswald and Golueke said, and this made systems like theirs vulnerable. One of the "chief disadvantages" was "the skilled attention which the space man must give to his cultures. His very existence will depend upon his 'green thumb' [in] manipulating the complex and highly sensitive physical-chemical systems proposed for recycling essential elements."[49] Apparently, they were not sure how well astronauts could handle these tasks. Thus, several years before ecologists started to think about Earth as a spaceship and investigated cycles of matter and transfer of energy, sanitary engineers thought of Earth as a closed system stabilized by interconnected loops. Those same engineers questioned the central importance of humans within a life-support system as well as the general ability of humans to handle the necessary task. Significantly, they arrived at these conclusions by thinking about human waste management in heaven and on Earth.[50]

Oswald and Golueke were not alone in trying to make bioregenerative algae systems work. NASA's largest contractors all had joined in, but many smaller companies, laboratory groups, and individual scientists also took to developing bioregenerative systems. Furthermore, the use of algae for life support in space

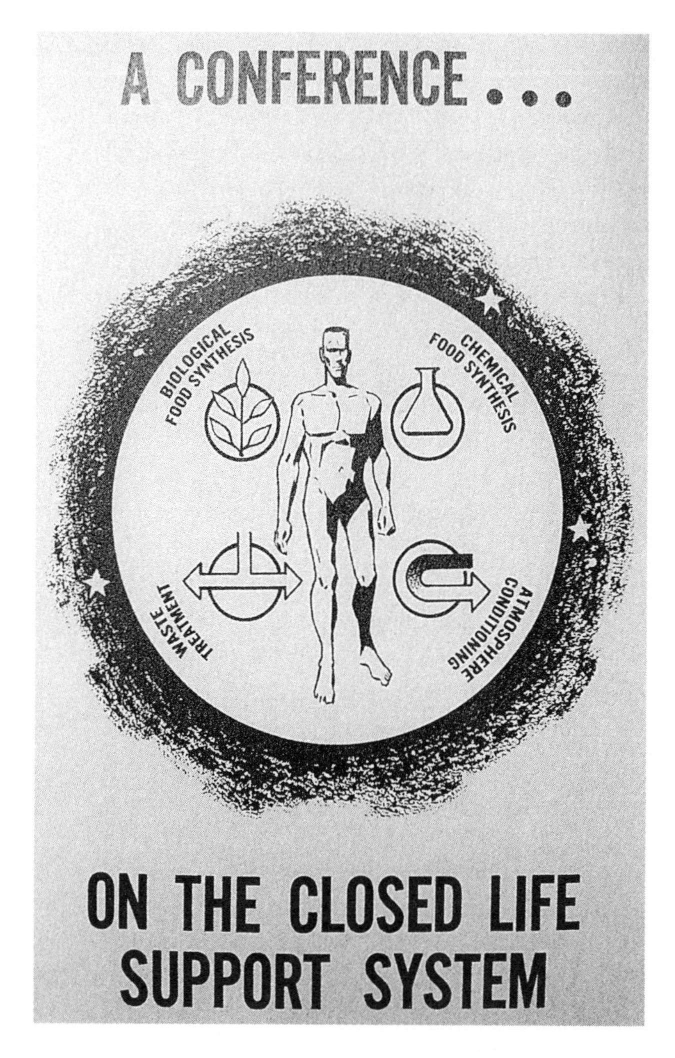

Fig. 2.5. Welcome pamphlet for a conference on the Closed Life Support System, April 14–15, 1966. Folder "Dr. Schnitzer Material on Long Term Life Support Prob.," Box 2, Series 36: Life Sciences Directorate, 1963–67, RG255, NARA.

was not confined to one side of the Iron Curtain; looking back on those years, Oswald once remarked that "a kind of 'algae space race' developed between the U.S. and the U.S.S.R."[51] Early on the Soviet Union pushed the vision of bioregenerative systems much further than the United States, up to the development of a fully functioning ecological system called the BIOS-3, which completed a successful test run in 1965 (see chapter 3).[52] Although it is mostly forgotten by now, at the time the subject matter was highly popular.

At least a dozen national and international meetings addressed the various issues related to waste recycling in space throughout the 1960s.[53] The topics were wide ranging. At a 1964 meeting at the University of Florida on Nutrition in Space and Related Waste Problems, for example, the audience heard life scientist Charles Ott describe Boeing's Managed Environmental System Assessment life-support system, which was designed for 150 days in space and processed all wastes via a "biological culture." This second attempt by Boeing at a complete life-support system dwelled at length on the many and varied types of wastes a human in a habitat might produce, including "(1) urine, (2) normal feces, (3) atmospheric moisture, (4) vomitus, (5) diarrhetic feces, and (6) personal hygiene waters." In a definite improvement on Boeing's first life-support system, Ott reported that "although this system is not at the flight hardware stage, it did provide a solid foundation upon which further engineering design and research and development may be based."[54]

Equally notable was the contribution at the same 1964 conference of Robert Krauss, a plant physiologist from the University of Maryland who introduced his own device for photosynthetic gas exchange aboard a spacecraft, which he called the Recyclostat.[55] Smaller than the Algatron, the Recyclostat was limited to processing carbon dioxide from the closed atmosphere of the space vehicle and turning it into oxygen through a continuously growing algae solution. It incorporated means to keep the air and water sterile, monitor the rate of exchange through the algae, and automatically harvest dead algae. Krauss's report is also interesting for his reflections on collaborative work on bioregenerative systems, which are similar to the NASA reports discussed in chapter 1. The critical problem, Krauss thought, was that biologists and engineers held different expectations and worked in different ways. Biologists, he said, were constantly frustrated at the engineers' "inability to understand the complexity and variability of the organisms which are to become parts of this operational system." At the same time, engineers were being "asked to provide hardware to support organisms about which our knowledge is remarkably primitive;" to them, this sounded almost like a contradiction in terms.[56]

Alongside bioregenerative systems, strictly physicochemical systems were also suggested, usually on the premise of unlimited availability of power. At the same conference in 1964 where the Algatron and the Recyclostat were presented, the audience heard famed nuclear physicist Edward Teller advertise a completely different approach. Teller clearly had no interest in algae, human waste, or ecological cycles. He rather argued that water could be created out of moon rock, with energy supplied by "a powerful nuclear reactor" on which "no amount of money should be spared."[57] Having this suggestion on the table alongside the algae system impressively documents the wide range of perspectives that were present in space science during these years.

ALGAL COMPLICATIONS AND THE TRIUMPH OF THE FECAL BAG

One should not be led to think, however, that the algae systems were necessarily easier to work with than physicochemical solutions; quite the contrary. One of the thornier issues, for example, was the subject of light. Within the inner solar system, sunlight was plentiful at all wavelengths, far beyond the point of photosynthetic saturation (it would only diminish as longer voyages headed outwards). However, given the demands of gravity, a single surface of illumination, and a maximum thickness of algal culture, the light source was logically best situated in the center of the Algatron's rotating drum. This implied the employment of an artificial rather than natural light source. The great advantage of using artificial light was that it could be controlled in terms of intensity and wavelength profile; the great drawback, of course, was that this was an extra drain on the ship's power supply. The plans called for an independent collar of solar cells to be deployed, but providing power for multiple environmental systems was beyond the capabilities of solar cell technology of the 1960s.[58] (Interestingly, unlike other spacefaring dreamers, including Teller, Oswald and Golueke did not fall back on nuclear power as a near-universal solution.) The engineers eventually favored direct or reflected sunlight as the source of light, incorporating water filters in "apertures in the chamber walls" to admit light into the two Algatrons via diffusers. These were installed as elements of temperature control: although the algae grew well under full sunlight, the cells were always at risk of suffering as they became too hot. Even more problematic was the insight that started to dawn upon scientists and engineers over the course of the 1960s: that, without extensive processing, the algae were almost indigestible for humans. Furthermore, the envisaged algae diet also turned out to be not particularly wholesome, as it contained too high a proportion of protein, but, at the same time, too few of the essential amino acids that humans needed to consume. This all was in line with the "aesthetic objections," as Oswald and Golueke called them, coming from astronauts who were disturbed by the thought of depending on their own recycled waste air, water, and perhaps even nutrition.[59]

For all these reasons, technological and otherwise, the system did not work properly by 1966, when NASA's space station ambitions were first trimmed. It would never be perfected beyond this point. The United States decided to fundamentally change the direction of its space program. In 1967 a NASA administrator offered to talk to the Congressional Committee on Science and Astronautics "in considerable detail in advance of most of our flight programs," but he turned out to be "rather reluctant to talk in very official terms about our possible long-run goals such as manned flight to the planets or lunar bases." By then, it had apparently become official policy to avoid pronouncements about long-term ambitions. In 1968 the last-generation prototype of the Algatron was tested; interestingly, it was then referred to as a "photosynthetic reactor," probably in view of the rise of nuclear power stations all over the country.[60]

In effect, systems of bioregeneration were never seriously implemented in early manned spacecraft, and the fecal bag made history by accompanying Apollo. Given that the lunar mission would only last about ten days anyway, this decision was perfectly rational at the time. In view of longer-term developments, however, it was perhaps unfortunate, and while limited work on closed life-support systems did continue until NASA geared up for space station *Freedom* in 1983, the work showed little of its former enthusiasm. With the Algatron, one may want to say with a bit of pathos, life, not just man, would have stepped onto the moon. There is no question that the Algatron's creators saw their device in such ways: Oswald said plainly that humans remained an unsustainable species on Earth without a bulwark industry of controlled photosynthetic growth of algae for waste purification, nutrient recovery, and food production. The use of algae was "the solution," Oswald wrote in the *American Journal of Public Health*, "of a few of the more interesting and significant problems facing civilization."[61]

3

THE PEOPLE'S "PLANETSHIP"

The toilet serves as the entrance to the whole [BIOS-3] complex.

—Iosef Gitelson, Genry Lisovsky, and Robert MacElroy, 2003

IN THE TWENTY YEARS AFTER WORLD WAR II, THE SOVIET UNION EMERGED as a technological superpower. Despite housing shortages and long lines for basic goods, they rattled the Americans with the testing of their own atomic bomb in 1949, far earlier than their adversaries had expected. A thermonuclear bomb followed in 1953, a mere two years after the Americans' test.[1] Perhaps the greatest Soviet triumph, however, came when the world heard an openly broadcast steady beeping from an orbiting eighty-three-kilogram sphere named Sputnik, launched on October 4, 1957.[2] Sputnik signaled to the United States, as well as to the rest of the world, the victory of socialist modernization and secularism.[3] After the dissolution of the Soviet Union in 1991, it has become increasingly hard to imagine how fundamentally the assumptions of people in the United States were altered when the Soviet Union—*not* the Americans—put the world's first artificial satellite into space. Sputnik took the Cold War straight to the heart of American fears. On the one hand, the satellite required a powerful and accurate rocket that could just as well carry an atomic bomb. On the other, the display of this technological feat undercut the self-assured American expectation of technoscientific leadership in the early Cold War.[4]

Like that of the United States, the Soviet rocket program was intimately connected to the ballistic missile development program: the exploration of space was loaded with very Earth-bound interests in ground surveillance and military expansion. Neither missiles nor espionage was amenable to becoming a pawn in the public relations game that unfolded during the decades of the early Cold War, so the matter was more or less kept secret. In contrast, the manned exploration of space was made highly visible and publicly celebrated. This to some extent parallels the contrast between the hidden development of nuclear weapons and the promotion of nuclear power plants on either side of the Iron Curtain.[5] Both areas, civil nuclear power and manned spaceflight, became prominent sites of technological collaboration between the otherwise competing superpowers.[6] In particular, historian James Harford claimed, space life science "was one of the few disciplines in which there was a relatively frequent and productive exchange of data between the Soviets and the Americans."[7] That exchange was facilitated by international conferences but more deeply by near-continuous overt intelligence gathering of published papers and their subsequent translation and distribution to the life scientists on both side of the Iron Curtain. Several memoirs of cosmonauts also vividly described the personal experience of living inside life-support facilities and on board the various Soviet space stations.[8]

Yet almost from the outset, both sides developed distinct priorities for their public efforts in space. Already around 1960 the Soviet program relegated a crewed lunar expedition to a time *after* interplanetary flights.[9] The Americans originally adopted the same timetable, but the sequence was famously reversed in 1962, when the goal of landing a man on the moon was declared the necessary first stage of interplanetary explorations. The differing priorities shaped the perception of what constituted success in space for decades afterwards. To the Americans, it was going somewhere and getting back; to the Soviets, it was going somewhere and staying there. As historian Asif Siddiqi rightfully argued, the Soviet space program must be understood on its own terms, particularly considering how the programs envisaged man's eventual presence in space. While the Americans were happy to celebrate their space program as a postwar achievement, Soviet space scientists traced their roots back to the ideas of the pioneer of rocketry and astronautics, Konstantin Tsiolkovsky.[10] Tsiolkovsky was a self-educated scientist and engineer who, around 1900, pursued philosophical concepts of human immortality in space. But he also developed influential designs of airplanes and drafted visions of space colonies complete with closed biological systems.[11] The goal of living in space was thus part of the origin story of and rationale for the Soviet Union's pursuit of space science and technology.[12]

Soviet ambitions in space charted a steady course progressing toward the goal of long-term missions. Between 1971 and 1985 (when the American space shuttle program was in regular service), Soviet cosmonauts amassed 3,300 person-days in space: an impressive record compared to the Americans' paltry 700 person-days.[13] To former NASA administrator Thomas Paine, it was clear that the Americans' lunar landings had barely dinted the "expressed determination" of the Soviet Union to "move out vigorously to colonize the solar system."[14] As Soviet cosmonaut Valentin Lebedev noted, after setting a new record in 1982 by spending 211 days in space, the Americans "don't really *live* in space—since most missions are only a week long, they merely tolerate *being* in space."[15]

Lebedev's sentiments were fully in line with official positions. In 1971, two years after the lunar landings of the Apollo mission, the assembled leadership of Soviet space life sciences published a sweeping programmatic statement—and they did so, pointedly, in English. Entitled "Theoretical and Experimental Decisions in the Creation of an Artificial Ecosystem for Human Support in Space," it stated, "all of man's former flights were not real ventures into space in the biological sense, as his life was supported with undegenerated earth supplies."[16] As late as 1986 major speeches in the Soviet world still carried the theme that "for space to be placed at the service of the people, we must learn how to live in it."[17] As a consequence of this consistent framing of their program, Soviet space scientists and engineers went far further in the development of artificial environments than their American counterparts did. They did so most prominently at the BIOS facilities at Krasnoyarsk, Siberia, as will be explored in this chapter.

Our aim in tracing these projects is comparatively modest. We provide further evidence that research in artificial environments was not an idiosyncrasy pursued by some mavericks from Berkeley but was rather a central concern of the most powerful space programs at the time. In parallel with their American counterparts, Soviet space scientists and engineers turned to study intensively the functioning of ecological systems and the use of algae and higher plants for the development of bioregenerative environments, and Soviet efforts were more comprehensive and often earlier than the Americans' attempts. Given our linguistic limitations, however, it is clear that this chapter can only be a first step toward a history of the experimental regimes that were headed at the development of bioregenerative life-support systems in the Soviet Union.[18]

LEARNING TO LIVE IN SPACE

With the doomed but highly publicized trip of the dog Laika in 1957, Soviet space scientists boldly advertised a functioning life-support system. Only three years later, in 1960, a pair of dogs, Belka and Strelka, became the first animals launched into space and successfully recovered.[19] Several smaller rocket missions followed that transported a wide range of organisms and living material into space and back, including ever more dogs but also mice, guinea pigs, reptiles, human blood samples, cancer samples, and seeds.[20] In 1961, finally, Yuri Gagarin famously became the first human in orbit. His flight was enthusiastically celebrated in the Soviet Union, even though, like the mice and dogs before him, Gagarin was mostly a passenger with no active role aboard—the spacecraft was largely automated, with centralized control firmly on the ground.

These launches were the beginning of an extensive research program in the Soviet Union to understand life in space. After 1962 a series of experiments investigated the effects of radiation and of prolonged weightlessness on animals, plants, and humans. Cosmonaut Gherman Titov is mostly remembered as the second Soviet spacefarer, and the fourth human in space altogether. But Titov is also known for a number of memorable firsts. Since he was only twenty-five years old when he went into orbit, Titov holds the record of the youngest person to go into space while he was also, on a less glamorous note, the first person to vomit in space. As Titov remembered, throughout the journey he suffered a bad spell of what later came to be called the "space adaptation syndrome," or simply "space sickness."[21] This predicament may have contributed to his lack of spiritual feelings in space, which was a common trope among American astronauts. Unlike his American counterparts, who often cast their journeys in religious terms, Titov assured an audience when he visited the United States that he had seen "no God or angels."[22]

Titov's visceral experience in space has been notable because Titov starred as one of the test persons of Soviet bioastronautics both before and after his flight—including research in the effect of extreme centrifugal forces, lack of oxygen, and sudden loss of atmospheric pressure. In these projects, bioastronautics was able to learn from the more established field of aviation medicine that also had taken to investigate human responses to extreme environmental strains. And these strains did not only concern the physiological condition of the body. As planes flew higher and longer and submarines traveled deeper and farther, studies abounded that attempted to judge the mental breaking point of people who receive only limited sensory inputs in confined spaces—psychology joined physiology in the space race. The endurance of long periods in orbit was not yet a problem in the 1960s because the first flights were comparatively short. Yuri Gagarin's journey lasted 108 minutes altogether, Alan Shepard's sub-orbital trip of the same year just a quarter of an hour. Nevertheless, they certainly represented extremes of isolation and limited space. In order to minimize the risk, Gagarin, Shepard, and all astronauts thereafter were kept busy throughout their journeys with constant checks, tests, and experiments, preventing them from dwelling too much on their situation. The stakes were obviously high: it was widely agreed that mental problems in spacecraft could ruin entire missions. The Cold War put not only the competing nations' technology but also their emotional resilience to the test. But before the cosmonauts could even develop psychological problems, they were in need of an adequate environment to survive in the first place. That required comprehensive ecological studies, in heaven as on Earth and, most importantly, in between. As Titov later explained, "no spaceman . . . truly leaves the earth. He must carry vital portions of it with him."[23]

THE BIOS FACILITY

In order to promote these lines of research, in 1961 a new institute was founded outside Krasnoyarsk, in the middle of Siberia, with the famed rocket engineer Sergei P. Korolev as its first director. At this time, Korolev had already become the celebrated chief designer of the Soviet space program, although his early career was far from smooth. In the course of the Stalinist purges, Korolev was arrested on false charges in 1938 but miraculously survived six years of imprisonment, including some months at the notoriously gruesome Kolyma labor camp. Korolev headed the Krasnoyarsk institute until his sudden death in 1966, when the helm was taken by the physicist-turned-biologist Boris G. Kovrov.[24]

Krasnoyarsk became a site for life science research within the Institute of Biophysics of the Soviet Academy of Sciences. There an ambitious program was initiated to develop artificial environments for survival during long spaceflights that soon became internationally renowned. Already in 1962, when UNESCO organized the First International Symposium on Basic Environmental Problems of Man in Space in Paris, it welcomed contributions from the United States and the Soviet Union alongside a raft of speakers from other countries. Norair M. Sisakian, a leading Soviet academician, delivered the opening address and set the priorities: "One of the most complicated tasks of space biology [that] I believe is the most magnificent task of contemporary natural science [is] the creation of a closed cycle for transforming matter in a spaceship cabin."[25]

This complicated and magnificent task, quoting Sisakian, fell to the scientists of the Krasnoyarsk institute including the physicist Leonid V. Kirensky, biophysicist Ivan A. Terskov, and photobiologist Iosef I. Gitelson. The centerpiece of the institute became the BIOS facility for research on closed ecological systems. Comprehensive work had to be delayed until the intercontinental ballistic missile program was assured, but by 1964 Soviet scientists and engineers had built their first technically sophisticated life-support system, the BIOS-1. Over the next two decades, layers of increasing biological and technological complexity were added, which gave rise to the more ambitious successor facilities BIOS-2 and BIOS-3.

The grand aim was to create a whole "planetship ecological system," as biologist (and aspiring cosmonaut) Andrei Bozhko later described. As we shall see, Bozhko became one of the test persons who spent a full year inside the BIOS-2. In this planetship, Bozhko explained retrospectively in 1975, many different types of organisms would eventually be included—"yeast, fungi, aquatic snails, slugs, fish, rabbits, chickens, etc."—in order to replicate near-natural cycles of matter and chains of food. "Plants and algae in this chain will be eaten by fish or other animals, which, in turn, can be used as food by the spacecraft crew," Bozhko said. But Bozhko's experience had taught him how difficult it was to build these artificial environments: a real planetship, he said, "is still far in the future, although, in principle, the feasibility of such projects is not in doubt."[26] In view of today's persistent difficulties in this respect, he may have been a little too optimistic.

According to the later description by Gitelson, the Krasnoyarsk institute was able to build on earlier work at the Institute of Biomedical Problems in Moscow, while it quickly developed its own research profile. Among the early successes in Krasnoyarsk were some daring experiments on human test persons. In 1962 one of the scientists, Yevgeny Y. Shepelev, enclosed himself within a sealed chamber and spent twenty-four hours therein together with several generations of algae cultures. As Shepelev observed afterwards, echoing some of NASA's earlier claims, in these ecological systems "man becomes a constituent, one of its functional links."[27] In fact, Shepelev's experience with algae essentially repeated the accomplishment of Boeing bioscientist R. H. Lowry from two years earlier, though Lowry's stay was only for eight hours.

In their use of algae as a vital component of bioregenerative environments, Soviet and American life scientists were in agreement. Indeed, the approach at Krasnoyarsk was remarkably similar to that of William Oswald's and Clarence Golueke's Algatron, though the Soviets were pushing these efforts substantially further to improve the system's efficiency. Gitelson, Kovrov, and others systematically tried to determine the conditions of maximum growth of chlorella algae. Using classic methods of plant physiology, the team exposed chlorella suspensions of various densities to white and monochromatic illumination, all the while holding the temperature, layer thickness, pH, and carbon-dioxide exposure constant. They demonstrated that while illuminating algae suspensions with stronger light intensities did produce rapid cell growth, it soon resulted in a maximum density, as the algae cells were blocking each other from incident light, thus inhibiting further growth. The team noted that "distinctly different illumination levels facilitated maximum productivity," namely 36–38 grams (dry weight) per liter over twenty-four hours, and they speculated that with more "complicated shapes" for the suspension vessel than simple rectangles, perhaps with multiple sources of light from many sides, they might gain higher yields in the future. They did not explicitly specify in these publications why the space program required growing chlorella in high yields, but other papers in the same published collection, *Problems of Space Biology*, noted that chlorella cells were incredibly useful for the basic components of life-support systems. They were able to deal with increasing carbon dioxide concentrations, which very quickly rise among breathing organisms in a closed chamber. At the same time, they were able to contribute to the removal of "organic wastes," which, as the pertinent publications explained, could be "mineralized in a biological filter by microorganisms."[28]

In the BIOS-1, the early sealed algae chamber was expanded beyond the basic two-element (man and algae) design that Shepelev had tested. It still used unicellular green algae to recycle water and air, but with a greater range of controlled environmental variables. BIOS-1 comprised a 12-cubic-meter chamber that was supplied with oxygen from an 18-liter algae cultivator illuminated by three 6 kilowatt xenon lamps, all designed to keep one person alive over a period of some days—successfully, as human experiments demonstrated over the next years. In the first of these trials, in 1964, Gitelson himself spent twelve hours inside the BIOS-1 facility, and later that year, another biologist at Krasnoyarsk, Yury Gurevich, stayed inside for a full twenty-four hours. Through these trial runs, the BIOS-1 had "experimentally" demonstrated, Soviet researchers claimed, that an "atmosphere for human respiration can be maintained with the help of continuous algal cultivation by our method, and the atmospheric O_2 and CO_2 concentrations can be kept constant." Gitelson's and Gurevich's enclosure established that "human and microalgae (*Chlorella vulgaris*)" were "biologically compatible with regard to gas exchange."[29]

The BIOS-1 facility demonstrated through trials of increasing length (five, fourteen, and thirty days) that humans and algae could in fact survive on recycled air from each other. The human test subjects came out of the chamber alive and, according to numerous tests, without noticeable physiological changes; but the algae also did well, which was good news. It was a much debated (and crucially important) question whether algae cultures were able to survive extended periods in closure without contamination and in stable equilibrium conditions—the answer was that they did. Over a forty-five-day experiment in which more than a hundred generations of algal cells succeeded each other, Soviet scientists concluded, "there was no tendency towards the deterioration of photosynthesis."[30]

It was later estimated that the BIOS-1 chamber initially operated only at about 20 percent closure, but even that placed the facility far beyond contemporary

American projects—not only in terms of survival period but also in terms of knowledge about potential sources of danger. In limited but important experiments, the Soviet team tested for "the accumulation of deleterious contaminants in dangerous quantities" in the cabin atmosphere—possibly knowing about the disastrous 1963 Boeing test run in the United States.[31] In the confinements of a space capsule, even small amounts of volatile impurities could become the cause of serious problems, and it was clear that any number of liquid or gaseous substances, which had never been tested for their effects, existed in miniscule quantities within these environments. On Earth, in more spacious premises that were in constant exchange of air and water with their surroundings, this never was a problem: stray substances were quickly diluted and processed, mostly through bacterial metabolism. But none of this could be taken for granted in space, where artificial devices had to make up for all natural exchanges, links, and cycles. As Shepelev explained in 1965, for this reason artificial environments would always contain "larger quantities of impurities than the terrestrial atmosphere," which implied that the tolerable limits of humans or other organisms had to be empirically determined. They needed to know not only the optimum values of environmental parameters "but also the permissible limits of transient and prolonged deviation from comfort indices."[32]

The parameters to be explored included potentially toxic substances that originated from the technical components of the system, such as corroded rubber tubes or cable sleeves. The life scientists also needed to look into issues of a more intimate nature, however. As V. V. Levashov commented in 1965, the "new conditions [of] manned space flight" necessitated rethinking the issue of personal hygiene, which was then still based, he thought, on textbooks and manuals that had been "virtually unchanged for the past hundred years." While hygiene on Earth was important enough, the concerns of "personal hygiene in space are at a wholly different level." It was a remarkable fact, Levashov quoted from "Soviet and foreign literature," that "halting or shortening normal toilet routine impairs the sense of well-being by the end of the second week . . . and causes skin changes or even diseases."[33] In small, enclosed spaces, Levashov explained, the interaction of microbiological, bacteriological, physiological, and environmental conditions dramatically destabilized the normal expectations of health. As a first step, Levashov himself focused on the reactions of his subjects' skin to those spaces, encompassing everything from studies of microflora to bedding.

However, a full exploration of these and other challenges demanded a more complex research infrastructure. BIOS-1 was a good start but definitely had its limits, as it was mainly designed to regenerate and monitor the constitution of the air. Food and water still had to be taken in and, though unstated, waste presumably removed. But the future cosmonauts in space could not survive by breathing only, and neither would they be able to live exclusively on algae, even if they wanted to. In the next generation facility, the BIOS-2, the Krasnoyarsk team therefore attempted to include higher plants as the basis of sustainable nutrition. And they also pushed the trial period up of humans inside an artificial environment to a new all-time record.

A YEAR IN A STARSHIP ON EARTH

On November 5, 1967, the medical doctor Hermann Manovtsev, technical expert Boris Ulybyshev, and biologist Andrei Bozhko walked through the steel door of BIOS-2 to begin a yearlong experiment inside the facility. It was supposed to trial important aspects of Korolev's last great vision, the Heavy Interplanetary Ship, which was envisaged to travel through space for years or even decades.[34] The central question for the BIOS-2 was whether the three participants would be able to survive unharmed in body and mind within the confined space. Another, subordinate aim was to see how far automation could be pushed in the growing of food in a space ship, and how a system of monitoring and control could help to maintain life throughout its interplanetary journey. The endeavor was kept strictly secret; even the participants' families were told that Manovtsev, Ulybyshev, and Bozhko were going on a test run in the Arctic. Only several years later did the story emerge, when Bozhko described his experiences inside the BIOS-2 in his 1975 memoir, *A Year in a "Starship."*

BIOS-2 moved the Soviet development of closed life-support facilities into its second stage. While BIOS-1 achieved in its final configuration an impressive 80 percent closure of the environment by having a water recycling system in place, it had become clear that to render the system fully self-sustained, the algae had to be complemented with other, more edible plants. A second chamber was therefore attached and encompassed some twenty square meters of floor space for hydroponically growing food crops. In this second chamber, a variety of species was cultivated to ensure the inhabitant's nutrition—potatoes, beetroots, carrots, cucumbers, and, in view of its digestive qualities, dill. Even wheat, a rather demanding crop, was tentatively included. These plants also contributed to regenerating oxygen to the atmosphere, though three-quarters of this important function was still achieved through algae cultures.[35] The greenhouse chamber was referred to, fittingly, as a "phytotron," thereby linking it to the controlled greenhouse facilities that were sprouting all over the world at the time.[36] Like other phytotrons, BIOS-2 incorporated fully controlled air-conditioning systems and water supply, while in contrast to others it was entirely run on artificial light: "Our 'sun' glows in the form of twelve xenon lamps," Bozhko later remarked.[37]

The use of artificial light increased the energy requirement of the facility, but the great advantage was that it enabled experiments even with those parameters that are usually fixed. For any location on Earth, for example, the sun determines the length of day throughout the year, and the incidence of light changes with weather conditions. None of this was inevitable in a phytotron. Under xenon lamps, the immutable "day" suddenly became fungible, while quality and quantity of light incidence could be set to any desired level. Quite explicitly, the BIOS-2 experiment did not run on Earth days but on lunar days, which consists of 360-hour cycles (15 Earth days) rather than 24-hour cycles. Apparently, plants adapted well to these conditions, and Bozhko recalled that a certain rhythm was developed by the crew in order to ensure constant supply: "A 'day' in our greenhouse lasts fourteen Earth days. Then 'night' begins, which lasts just as long. . . . In order for [the plants] to accumulate biomass, we conduct our sowing at night, since seeds germinated thusly do not need light. When our 'sun' blazes, the plants meet it with already unfolded leaves. The sowing of seeds and the harvesting are done at different times periodically by way of a 'conveyor' so that there would always be fresh greens at the table."[38]

The circadian and seasonal rhythms of plants and people, specifically their durations and malleability, became a focus of the experiments. Up to 1965, little work had been done on such questions, but the prospect of being exposed to completely different changes of day and night on space stations or lunar bases made scientists reconsider the importance of these issues. Future spacefarers might have to adjust to lunar or Martian time schedules. Perhaps even their perception of time could be changed: at some point, Soviet scientists entertained the idea to have spacefarers enact twenty-eight-hour days, so that "the feeling of a '4-year' flight" would in fact last five Earth years.[39]

We can take from Bozhko's description that the BIOS-2 "starship on Earth" consisted of a laboratory and a habitation module, imagined to be part of a long thin spacecraft design and complemented by an electric rocket engine. Electrical power was, in fact, supplied externally to the BIOS facility. The whole living compartment measured four meters in length and three meters in width; it had three bunk beds and small folding tables. Bozhko told his readers how, in a "tiny galley," he and his two crewmates cooked food, while also cultivating future meals in the attached conservatory. The fresh food supply was only an addendum, however. Although by then three phytotrons had been added to the system, the crew still lived mostly on canned food, with a calculated limit of 1,000 calories per day (by comparison, today's astronauts consume three times as much in space). Finally, there was a washing allowance of one bucket full of water per every ten days—provided the recycling system was working. The endurance of confinement, hunger, and austerity was obviously part of the setup.

PERSONAL FITNESS AND GROUP SHOWERS

A year of life in a closed room was "no joke," Bozhko said, which was probably a tremendous understatement. A human drama took place inside the chamber whose occupants were cut off from outside human contact for a year, unable to see or hear anything from friends and family, and unable to escape each other's presence. "It examines a person," Bozhko said. Part of the BIOS-2 trials was to expand the medical and physiological tests of humans in artificial environments by evaluating their psychological response to extended isolation and confinement. Some of the challenges were already known from the existing cosmonaut program, as well as aviation and submariner studies as mentioned earlier. The leaders of the program, however, had been unable to test the complex group dynamics so far. Bozhko recalled how the Soviet psychiatrist and psychologist Fedor Gorbov, who was specifically concerned with space psychology from Gagarin's flight onwards, commented upon this situation: "Even knowing well the characteristics of each member of the group, you cannot predict how the group will manifest itself as a whole. What kind of relationships will develop between its individual participants? A group is not the arithmetic sum of individuals, rather a new, single organism with its own laws of development and life." Bozhko was very clear that this was the hardest part of the experiment. As he quoted from O. Henry's "Handbook of Hymen": "If you want to instigate the art of manslaughter just shut two men up in an eighteen by twenty-foot cabin for a month. Human nature won't stand it."[40]

According to Gorbov's published work that made it to NASA, the BIOS facility offered an interesting situation to test for the mental stability of cosmonauts. Gorbov had long used a number of psychological tests to gauge who was "fit for spaceflight." As he explained in a volume produced by the Soviet Academy of Sciences in 1966, "the first step" to this evaluation was the "isolation of the factors of psychological well-being and the development of suitable methods of investigation." These factors, according to Gorbov, included rest and sleep, memory, and physiological functions, which for each individual could be monitored over time. But now they were moving on to test the suitability of groups for future "protracted spaceflights." To this purpose, Gorbov and his collaborator Mikhail A. Novikov relied on "homeostatic" methods originally developed by the British cybernetic psychologist William Ross Ashby. The contrived situation in Gorbov's version was a shower room. Four men enter a shower room fitted with pipes of varyingly interconnected temperature and pressure. Any man could choose his optimal shower comfort but only at the expense of his fellows. "Only at the price of refraining from egocentric behavior was it possible to work out a set of conditions acceptable to all, which required a 'conflict' of strategies in the group situation," explained Gorbov and Novikov about the rationale of their experiment.[41] Indeed, they explicitly noted that the advantage of the shower-room experiment and its subsequent incarnations was that it identified the "leader" of the group, namely, the person who was able to negotiate a consensual solution. Gorbov took a keen interest in the BIOS-2 experiment as a first trial run of a more realistic rather than contrived situation to assess and measure the parameters of successful group dynamics.

The first two months of BIOS-2 were the hardest to endure: they simulated an actual flight, during which the crew remained enclosed in their living quarter. Only then the station "docked" with the greenhouse module. The opening of the phytotrons to the crew was a crucial moment. On the one hand, one of the central purposes of the experiment was to trial a functional growing space of plants for food. On the other, access to the greenhouse brought some psychological relief—not only because it hosted other living beings and provided aesthetic pleasure, as Bozhko later underlined, but also because it considerably widened the space and gave the participants opportunity to get further away from each other. Bozhko recalled his delight as they opened the phytotron to reveal the growing herbs and plants. The facility was productive, though never quite productive enough: it generated about four kilograms per square meter of kale, watercress, and fennel—a nice complement to the diet but by no means sufficient to keep three adults going. Bozhko therefore speculated, in hindsight, about how yeast protein might one day replace plants as a source of nutrition. The pertinent passage is worth quoting at length:

> According to many experts, greenhouses with higher plants in the long-term space mission are able to provide full-fledged astronauts carbohydrates, vitamins, mineral salts. . . . However, it is now possible to imagine another way to create a complete dietary supplement: . . . microbial or yeast protein. The fact is that the productivity of microorganisms is on many orders of magnitude higher than the productivity of animals and birds. Thus, for example, a ton of yeast is able to give a thousand tons of descendants, that is, up to 400 tons of protein. Already today, at the initial stage of research, from yeast protein dishes can be prepared pleasant in appearance, fragrant and delicious: Bouillon, jellied meat, etc. The dry yeast protein in normal conditions can be preserved indefinitely. It can be easily

> cleaned from impurities, has a pleasant taste and can serve as the basis for cooking. It can be obtained by way of mechanical or chemical degradation of the yeast cell membranes. . . . One gets a white tasteless powder, which, like any other pure anhydrous protein can be stored for a long time. From it you can cook delicious and aromatic dishes, to which it is sufficient to add a usual taste—sweet, sour, salty, bitter. . . . Perhaps, here might be the successful solution to the question of nourishing the cosmonauts?[42]

Even though the green room of growing plants offered some benefit, as this passage indicates, Bozhko worried about the reliability of continually recycled nutrients and, above all, water. Solid waste was not recycled inside the BIOS-2, while urine was. Their water, as Bozhko wrote in retrospect, was "repeatedly obtained from urine and other wastes of the vital activities of our organisms. And we had to drink only this water . . . to experience its effect on oneself."[43] Part of the experimental protocol was to test the feasibility, reliability, and health from drinking continuously recycled water just as the BIOS-1 trials tested continuously recycled air. The water was, as far as anyone could tell, fine, but the suspicion that bio-regenerated products would have deleterious effects continued.

This was one reason the inhabitants were continuously monitored: it was important to test the properties of the room, but it was equally important to understand the performance of the human organism under enclosed conditions. Medical checks were scheduled every morning and evening; the participants even had to wear special helmets at night that recorded encephalograms. Every ten days, emergency situations were simulated, such as a dramatic rise of temperature or a substantial change in the concentration of oxygen and carbon dioxide in the chamber's atmosphere. In a nice turn of phrase, Bozhko realized that he and his team were both "the experimentalists" and "the rabbits." Years later, Bozhko still thought that the toughest were the "exchange days." The crew was required to record the exact amount of water they drank and urine they expelled for several days beforehand, and even ingest dye to mark their solid waste (which was then exported as medical samples). Blood was another part of this surveillance. It was extracted at regular intervals because its composition was considered a "metabolic mirror" that could signal possible disorders at an early stage.[44] It allowed tracing the humans' connection to their environment via macronutrients such as nitrogen and oxygen and micronutrients such as titanium, copper, and molybdenum.

By the last months of the BIOS-2 experiment, the growing and testing, medical exams, and trapped loneliness had become routine for the three men. While they stayed for the full year, it was in large part merely an exercise to display the conclusion to the world. In contrast to the gain of insight, the suffering had not decreased. "We cannot in any way get used to manage a very small amount of water in the washing and dishwashing," Bozhko bemoaned. "Every ten days is bath day; only ten liters of water per person. It is an amount of water just barely enough [to] rinse off the soap." The diet became monotonous, and one of the participants had lost three kilograms.[45] But they were released eventually, all of them alive.

Tragically, the three heroic test persons were never rewarded for their commitment. The plan had been that on June 8, 1971, they would become part of a Mars mission with the tested environmental system aboard.[46] However, a rocket exploded on the launch pad in 1970, and therefore all further engine and assembly development was delayed until the failure had been closely investigated. Departure was postponed to an unspecified date in the future—too long for the three BIOS-2 survivors to still be eligible.

Given Bozhko's memoir, it is interesting to compare the role that the biologists played inside BIOS-2 with that of cosmonauts on actual space flights and their respective perceptions. As historian Slava Gerovitch argued, the early cosmonaut cohorts were emblematically used to offer up a new vision of the autonomous individual citizen under Nikita Khrushchev—despite the fact that the actual role of the cosmonaut inside a space vehicle was distinctly limited.[47] The biological cosmonauts inside the BIOS-2, in contrast, were active components of the system as well as its adaptable self-experimenters, but in their sealed chambers, subsisting on self-grown potatoes and canned food hardly made them equally shining examples of the New Man in the Soviet space age.

CULMINATION OF EXPERIENCE IN BIOS-3

Over the following years, after the end of BIOS-2, the program expanded in two ways. First, Soviet cosmonauts began to take their plant experiments into space aboard *Salyut* Soviet space stations in the 1970s. Throughout the 1970s and 1980s, ground-based facilities informed space based research—and vice versa. One critical area was flowering. For nearly a decade, Soviet cosmonauts failed to generate flowers on the plants they took into space throughout the 1970s. Without flowering, of course, there can be no seeds, and thus no plant reproduction. Missions aboard the *Salyut-6* (1977–1981) and *Salyut-7* (1982–1986) were both charged with trying to find out why the plants refused to flower, even though some of them even developed buds. Lack of gravity was one of the prime candidate factors, so both missions were equipped with growth chambers and new experimental apparatus, including the Biogravistat, which placed a 1-g force on the growing plant, and the Magnitogravistat, which did the same with a heterogeneous magnetic field. Lastly, there was the Fiton, which contained five cells with individual light sources and seed placed into an agar substrate made up of 98 percent water. This eventually did the trick: in mid-July 1982, during cosmonaut Valentin Lebedev's record stay in space, the plants germinated and developed flowering buds, and by the end of August, Soviet scientists were able to celebrate some two hundred plants flowering for the first time in orbit. The ability to grow a plant from seed to maturity and in turn produce new seeds was, American James Oberg lamented once more, yet another Soviet "major breakthrough" in the space race.[48]

Second, the BIOS complex itself grew substantially between 1969 and 1974. The successor to the BIOS-2 was, inevitably, BIOS-3, which was a solid three-hundred-cubic-meter facility. From 1972 until 1984, three teams of three biological cosmonauts each stayed inside the BIOS-3 for six months apiece. Each team faced an altered experimental environment while growing food in the phytotron sections of the facility. All crops were grown in portions, with fourteen stages of growth in sequence, including dwarf wheat, carrots, beets, potatoes, onions and, curiously, tiger nuts (for producing oil). Each of the three phytotrons had a conveyor system; as soon as one portion of crop was harvested, the conveyor moved the next group forward. This system saved space, as the occupant only needed access to the front of the conveyor belt, and thus also kept the inhabitants better supplied with fresh vegetables throughout their stay. Apparently, the conveyor brought the system into a steady state and satisfied at least half of the crew's food requirements. Each participant received two hundred grams of grain for baking bread and four hundred grams of vegetables per day, complemented with canned meat for more protein.

BIOS-3 was also equipped with algae cultivators that recycled human wastes back to air and water. One major conclusion to emerge from the BIOS-3, however, stretched the Soviet experiments far beyond the Algatron and comparable American systems: "if microalgae are cultivated in a mixed culture, essentially in a naturally established relationship with bacteria rather than as axenic algal strains, the system becomes quite stable." Eventually, the system was almost self-sustaining: the Soviet scientists trumpeted a final figure of 94.5 percent closure of gas and water supply by the end of their experiments in the mid-1980s.[49]

Solid waste was mostly dried and some of it reused in the algae cultivators. Its collection vessel, however, "the toilet," was placed right at the entrance to the complex.[50] The environment was designed, so to speak, from the bottom up. Of course, once the external door was closed it would not be opened again until the completion of the trial, thus instantaneously converting the facilities' front door into its most remote component. Still, there was no room for aesthetic sensitivities about waste inside the BIOS-3 or the Soviet orbiting space stations, and the cosmonauts, like their American counterparts, faced the problem of sanitation with a stoic realism. Orbiting above the Earth in *Salyut* 7 on May 19, 1982, Lebedev began his diary entry, "at midnight right after we finished pumping urine." He openly admitted that "it wasn't a pleasant procedure," but not because, as he wrote, "we were messing with urine like sewage workers—after all, it's our own waste—but because the system has many different plugs, adaptors, and tubes that we have to connect and disconnect them many times."[51] It was the engineering systems that Lebedev found unpleasant, not the urine.

The first trial run of the BIOS-3 came in 1972. The aim was to experimentally establish the requirements of a minimal ecological system consisting of a suspension of chlorella algae plus bacteria, vegetables, and humans, in which the nutrient cycle for humans was closed, and in which nearly one-quarter of the oxygen returned. At the same time, the hope was that the large quantity of the volatile substances produced might be reciprocally absorbed or processed by organisms in the system. As in the BIOS-2, the "experimenters" were embedded in this system in multiple ways: they planted the crops in the phytotrons and harvested the outcome, but they also used their excretions—gaseous, liquid, and solid—to feed the algae cultures. All in all, this worked fairly well, as we learn from a summary statement looking back on the dozen years of experiments: "The sanitary water used (shower, wash-basin, etc.) after filtration, along with the urea of the experimentators without treatment or storage, was introduced into the wheat nutrient medium. Thus wheat's needs were met 95% in nitrogen, 28% in sulphur, 19% in potassium, 17% in phosphorus and 19% in magnesium." At the end of a two-month trial, the team concluded that a nine-vegetable ecosystem combined with an algae culture satisfied the nutrient requirements for a single person. It also provided, they claimed, an "improved psychological perception of the environmental conditions."[52]

The scientists continued testing a variety of photoperiods for space stations, lunar missions, and potentially extra-solar-system journeys. Several experiments even extended the periods of darkness beyond the fifteen days of BIOS-2, and also reduced the temperature to three degrees Celsius. It turned out that they had some successes with wheat and peas surviving the long night, but also failures, as tomatoes and cucumbers perished.[53] And there were some interesting experiments on complementing the vegetarian nutrition with meat. In this context, quails—of all animals—were reported as a potentially attractive source of animal protein. According to a Soviet space agency publication, "Quails in Space," quails surpassed either chickens or rabbits in both their reproduction rate and in their record-breaking speed of converting food into body weight. At the same time, quails generated some thirty times their own weight in eggs per year. Moreover, their eggs were "absolutely sterile," the report underlined: "no matter how long they are kept they never spoil." Thus, on the envisioned mini-farm in space, quails and their eggs were near-perfect food—although so far they have not made their way into a spacecraft or space station.[54] One unresolved question, again, might be the issue of waste management. Quail diapers hardly seem practicable.

Finally, BIOS-3 explicitly addressed the question of how long the crops grown in the phytotron would remain stable over extended periods. An interplanetary ship, and especially an intergenerational ship, needed to maintain the crops' properties as much as possible and try to arrest natural evolutionary processes.[55] Was it possible to build an artificial environment that would offer up plants with the same properties generation after generation?

These questions had been part of plant science for a long time, and Soviet scientists had been at the forefront of the pertaining investigations. Infamously, the Soviet agronomist Trofim Lysenko ruined both this field of research and Soviet agricultural productivity and encouraged the purge of Russian geneticists with his politically convenient theories of inheritance.[56] Lysenkoism was used on the other side of the Iron Curtain as a stark example of the dangers of imposing an ideology on science. In the United States, genetics became synonymous with anticommunism, while plant physiology became suspicious.[57] Lysenko's legacy hung heavy on Russian biology for more than a generation, but Lysenkoism was not emblematic of Soviet science as a whole.[58] It seems likely, in fact, that the priority assigned to the environmental impact on biological systems meant that Soviet biologists were excellently prepared to explore ecosystems for space habitation. Unlike in the Western countries, where plant physiology rapidly declined through the 1960s and 1970s, in the Soviet Union experimental work continued on a large scale.

THE SETTING OF THE SOVIET SPACE PROGRAM

More humans have survived inside the closed habitats of the BIOS facility than have walked on the moon. By 1983 the American Office of Technology Assessment regarded the Soviet space program as a steady and reliable feature of the international landscape and accepted as entirely realistic Moscow's reports about the Soviet desire to establish a permanent "orbital manned complex."[59] James Oberg, a major author on the space age, acknowledged at the Case for Mars II conference in 1984 that "a key technology development effort in Soviet space activity involves garden in space. [Cosmonauts] in orbit succeeded in growing plants from seed. [On] Earth, experimenters have set up hermetically-sealed closed-loop ecologies with humans inside, which have functioned for many months." By 1987 the Soviet *Mir* space station was in steady use, while there was hardly any progress on building an American (or international) space station, which prompted some to comment that the Soviet Union had once more moved "ahead of the U.S. in manned spaceflight." *Time* magazine reported the "Soviets Surging Ahead" with *Mir*.[60]

On the eve of the Soviet Union's collapse, the program to solve the biological problems of long-term space habitation and journeys continued, and Russia's workhorse satellite launch facilities and rockets outlived their American rivals. In the 2010s Russian rockets still were responsible for resupplying and delivering crews to the ISS.[61] But in neither *Skylab* nor *Mir* nor the ISS has any fully bioregenerative system been installed. In part, the technology was still not sufficiently developed, but another great hurdle, the leaders of the program concluded, was misguided expectations about the reality of living in space. "Unfortunately," Gitelson, Lisovsky and MacElroy wrote in 2003, "neither Soviet cosmonautics nor the astronautics of the USA has managed to overcome the justified conservatism of engineer thinking and the unjustified prejudice against 'incomprehensible' biological technology. It only remains for us to hope that the next generation of spacecraft designers must appreciate the advantages of biotechnological installations for human life support during long-distance and long-duration space flights in the twenty-first century. Our results are meant for them."[62]

GARDENS IN SPACE

ROM: A good waste extraction system is important. Imagine where we'd be without one.

QUARK: Of course it's important, but my point is, you're not getting the choice assignments.

—*Star Trek: Deep Space Nine*, season 5, episode 5, "The Assignment"

IN THE DECADE AFTER APOLLO, AMERICAN SCIENCE AND SOCIETY NEARED THE bottom of a deep melancholy. A pinnacle of prestige had been reached, and seemingly passed. The triumph of Apollo 11's landing on the moon in 1969 televised an image of America as an enlightened nation that strove for peace on Earth, but that picture was soon overwhelmed by the gruesome Vietnam conflict and increasingly violent race riots that dominated the news. Postwar science architect Vannevar Bush pleaded in his memoir of 1970 that "we need, today, something we can be genuinely proud of. . . . It should help dispel the gloom."[1] Instead, chemical and automobile companies were accused of putting the people's health at risk, while the corruption of the military-industrial complex was exposed by greedy, shortsighted decisions. And of course there was the Watergate scandal, which dealt the final blow to the Richard M. Nixon administration. All in all, it was "a kidney stone of a decade," a *Doonesbury* cartoon character toasted as the 1970s ended.[2]

In the course of all this, NASA's role on the national stage was much reduced, and not only because of economic recession. Having long contracted with the military-industrial complex, NASA came to be viewed by many as the ultimate expression of technocratic modernism, which eroded democracy in favor of command-and-control centralization. Many scientists dropped their declared political neutrality to rally against the use of science and technology for the military state. NASA tried to hit back against the ascendant counterculture, which it thought was to blame for its declining fortunes. The head of NASA, Administrator Thomas Paine, berated a graduating high school class in 1970 with a speech about the distinction between the influential popular culture of "Potland" and the boring but steady commitment to mathematics and reason of "Squareland," which the space program evinced. Yet this in no way addressed the charge that NASA so far had employed exclusively white males as astronauts.[3]

In view of its tarnished popular image, NASA launched the first experimental orbiting space station, *Skylab*, in May 1973 with an emphasis on science and medicine. It also featured the long-anticipated first space telescope to make observations above Earth's atmosphere. The crews of the three missions between May 1973 and February 1974 underwent extended physiological experimentation to learn about the parameters of their bodily comfort and maintenance. It is widely assumed that this was the first time the physiological and social aspects of living in space were investigated.[4] As previous chapters of this book have demonstrated, that assumption is untrue, but NASA undoubtedly attracted badly needed publicity in the mid-1970s by designing a shower and toilet for the orbiting astronauts. The space commode was widely announced in such publications as *National Geographic*. But NASA's own pronouncements adopted a distinct tone of understatement. The collected feces were vacuum-dried "and returned for analysis," *NASA Facts* assured, while the official NASA history of Skylab noted blandly that "the engineering problems involved in collecting liquid, separating it from air, measuring it, and accurately sampling it, all in zero g, were formidable."[5] NASA continued to tread a delicate line between picturing astronauts as larger-than-life heroes and rendering their life in space as normal as possible with bathrooms and treadmills.

Nothing stopped NASA's decline, however. The last two missions to *Skylab* were canceled in 1974 as the station's orbit gradually deteriorated. It eventually crashed to Earth on July 11, 1979. (Characteristic of the decade, even *Skylab*'s unplanned reentry was a failure. Aiming for the ocean surface south of Cape Town, hundreds of pieces of *Skylab* struck Western Australia instead.) In view of its political and institutional weakness, NASA commissioned a planning document as part of the budget process for 1976, entitled "Outlook for Space." According to this plan, NASA anticipated "no spectacular high cost technology development." NASA would instead send uncrewed exploratory missions like the Voyager probes to the outer solar system. They eventually returned stunning images of Jupiter and Saturn, but nobody was sure of those results at the time. In 1976, in fact, NASA administrator James Fletcher assured Glenn Schleede, then President Gerald R. Ford's domestic policy advisor, that "what is not seen as happening" was a "reinitiation of manned space flights to the moon," "the establishment of a lunar base/colony," or "initiation of a mission to land men on Mars." Fletcher's planning rationale was wisely chosen. Ford's conservative administration insisted that large federally funded programs be trimmed. The domestic policy team suggested, for example, that others might fund some of NASA's branches, such as commercial users of satellite technology.[6] It was only after Ford lost to Jimmy Carter in the close election of 1976 that NASA tried to revive some of its higher aims with the new space shuttle, and the planning of another space station after the reelection of Ronald Reagan in 1984.

Between the late 1970s and the early 1980s, the quest for artificial environments that supported life in space became more important than ever as NASA sought new frontiers. Opening the Case for Mars conference in 1984 in Boulder, Colorado, the former NASA administrator, Thomas Paine, announced to a huge audience that "when space navigation and closed-ecology life support technologies have been mastered, Mars will be settled."[7] Yet figures outside NASA also received wide popular attention. This chapter introduces some of these visionaries, such as physicist Gerard O'Neill, who designed cylindrical space colonies, though strikingly he did not spend any thought on the lowly question of these colonies' sustainable resource management. Describing O'Neill's vision helps us to put NASA's own projects into perspective. We then turn to NASA's different approach to bringing humanity into orbit, which restarted in 1979 under the leadership of Robert MacElroy and Maurice Averner and crystallized around a design that came to be called the Controlled Environment Life Support System (CELSS). It aimed at providing artificial environments for America's upcoming space station, emblematically named *Freedom*, but the Cold War ended and *Freedom* never materialized. Substantial parts of the preparatory work, however, were incorporated into major projects of the 1990s. One of the most significant will be described at the end of this chapter: the Lunar-Mars Life Support Test Project (LMLSTP), which prepared and tested some of the first modules for the ISS launched in 1998.

THE FICTION OF LOTS OF SEX AND NO WASTE

Much of NASA's work in this period was decidedly "square," to use both Paine's rhetoric and historian Cyrus Mody's term—solid; characterized by routine, "normal" science; and not fancy in any respect. Square science did not resonate well with much of the era, when conservative institutions as well as liberal leftwing groups and movements dominated the public.[8] In the mid-1970s, for example, a glamorous vision of living in space had emerged that was largely independent of NASA's own scientists and engineers and was definitely not square. That vision included ideas of interplanetary, if not intergalactic, travel and the creation of luxurious space settlements, regardless of the limits of technological feasibility.

Much of this exciting speculation built on the earlier work of physicist Freeman Dyson. Dyson rose to fame for his important contributions to quantum electrodynamics and solid-state physics, but between 1957 and 1961, Dyson worked for General Atomics on the Orion Project. The project aimed at developing spacecraft driven by "nuclear pulse propulsion" for interstellar journeys: Starship Orion would be pushed through space by exploding atomic bombs behind it. Dyson's design built on the 1950s promise of limitless atomic energy. His hope was to reach Mars by 1965 and Saturn by 1970. In the wake of the Partial Test Ban Treaty of 1963, the project was abandoned, but Dyson never stopped dreaming about journeys on interstellar space ships.[9]

He was soon joined by others. Physicists such as Gerard O'Neill and Eric Drexler picked up the thread and sold their utopian ideas to enthusiastic audiences in the later 1970s. In 1969 O'Neill had tried to join NASA's crew of astronauts and, when rejected, decided to satisfy his yearning for space by other means. Probably the most famous proposal came in 1974 when O'Neill developed and publicized strategies for human space habitation. He outlined space colonies in a specific design that came to be called O'Neill Cylinders: cylinder-shaped settlements, twenty miles long with a five-mile diameter, that rotated along their long axis to provide artificial gravity. These cylinders would be stationed at critically stable Lagrange points to keep them in a fixed position relative to Earth. They would gather energy from the sun via mirrors, and material via "mass-drivers" that flung rock from asteroids and the moon. Eric Drexler later further developed these ideas of O'Neill's (but primarily made himself known as one of the first nanoscientists).[10]

O'Neill and Drexler were increasingly frustrated by NASA's refusal to support their projects, even though their seminars and publications attracted huge audiences and devoted followers. Given their peculiar approach that combined utopian audacity with scientific and engineering competence, historian Patrick McCray termed them "visioneers." McCray argued that O'Neill's ideas in particular "responded to prevailing pessimism about technology and profound concerns about the deteriorating environment." One of O'Neill's contemporaries alternatively noted, however, that this advocacy of space colonies derived from the frustration of young university students with the Vietnam War and the misguided use of science and engineering in America at the time. The visions brought forward by Drexler, O'Neill, and Dyson were fully in line with high-flying attitudes of contemporary technocratic elites—there was still much enthusiasm in the 1970s about nuclear motors, railguns, and rocket engines.

These advanced boys' toys continued the array of masculinized, sexy technologies like cars and computers, which were so much more inspiring to fellow physicists, engineers, or even most lay citizens, than the day-to-day work of life support.[11]

The visioneers' glamorous plans largely ignored the mundane problems of living in space—such as life-support systems or the regeneration of resources. O'Neill and others were still stuck in the logic of the 1950s, where resources were apparently infinite.[12] This was true even of Drexler, who was highly critical of the official space program throughout the 1960s. Drexler spoke eloquently about the problems of a growing population on Earth, and his own solution preferred "inexpensive propulsion to supplement flashy, inefficient rockets."[13] Yet neither Drexler nor O'Neill embedded his technology into a sustainable approach to life in space—empty reservoirs would be replenished by mining the resources of asteroids. Thus, not only did they not contribute to alleviating the ecological and social problems on Earth, they carried these problems into space, completely oblivious to the implications. Nevertheless, these sketches of a possibly glorious future immensely appealed to an America still reeling from the 1973 energy crisis.

That visionary technocrats ignored the practicalities of living in space has been a theme explored in science fiction. In Kim Stanley Robinson's award-winning 1993 novel *Red Mars*, for example, geologist Ann Clayborne challenges engineer Nadia Cherneshevsky to move outside of her preoccupation with waste management: if she was going to be a "plumber" "installing *toilets*," Ann says, "why did you come to Mars after all?" Much the same message was used in the 1990s *Star Trek* spin-off, *Deep Space Nine*, which moved the show from a space-cowboy adventure to storylines of colonization and imperialism. In a humorous moment, one character noted that "waste extraction" was undoubtedly essential, but his brother retorted it also was not anyone's "choice assignment" in the engineering department. Similarly, in Andy Weir's second novel, *Artemis*, one main character asks a fellow lunar city inhabitant who tends the grounds, "Let me ask you something. . . . Who moves all the way to the moon just to mow lawns?"[14]

These science-fiction stories illustrate and satirize cultural assumptions about the value of various labor roles in the space age. Even in the 1970s, however, both the technologists and the environmentalists were criticized for ignoring the practicalities of space life. The creator of the famed *Whole Earth Catalog*, Stewart Brand, joined with journalist and ecological activist Peter Warshall to edit the book, *Space Colonies*, which reprinted O'Neill's testimony before Congress but also included his numerous critics.[15] Biologist Lynn Margulis, among others, was offered the chance to comment on O'Neill's conception of space colonies for Brand's book. Margulis considered one of the major weaknesses to be O'Neill's breezy assumptions about life support: the "John Todd's of the World (e.g. holistic biological thinkers and doers) must connect with O'Neill and his crew to help stop the handwaving. Many details are not easily worked out simply because it is said that they are easy [like the] delivery of all needs [and] the removal of all wastes." John Todd had built an eco-village that produced food by means of sustainable agriculture and aquaculture. When Todd also contributed to Brand's call, however, he revealed that his knowledge of NASA's closed system ecological work had stalled around 1962, when Howard Odum had charged that space biology was too simplistic and remained seemingly unaware of the decade in between.[16]

Margulis believed that visioneer physicists like Drexler, O'Neill, and Dyson brushed aside the issue of air, water, and waste recycling because they believed it was trivial and easily solved. Her assessment was correct. Dyson's Orion starship was planned as a vehicle with huge storage capacity and unrestricted access to energy and other resources. It would be able to haul such enormous amounts of freight that the painstaking economy of weight, usually the top priority of space travel, would be irrelevant. It could carry hundreds of tons of water and food, sufficient for opulent meals throughout the journey, and waste would be thoughtlessly discarded into the vastness of the universe. Dyson's spacefarers would coast along in leisure, he said, swimming in water and feasting on roast beef. Twenty years later, O'Neill similarly imagined a traveler through the inner solar system, "joined in his stateroom by his friends for a bon voyage champagne party." No wonder these visions of a land of plenty in space were more appealing to public audiences than the sober reality of astronauts drinking their own recycled sweat and urine. To Peter Warshall, however, such fantasies were "prime examples of contemporary American schizophrenia: a technological romanticism totally removed from agricultural practicality."[17]

There was, in fact, a near obsession among the visioneers with the pleasures of living abundantly in space. In his 1977 book on *Colonies in Space*, Thomas A. Heppenheimer, one of O'Neill's most significant acolytes, devoted an entire chapter to sex in zero gravity, while waste management was given a mere two pages. In Heppenheimer's imagination, glorious days would be spent in space swimming in zero-gravity pools suspended along the central axis of the cylinder. He dreamed of people flying through the low-gravity center with wings or motors. They would have in-home entertainment of stereos, tape decks, and TVs (he oddly assumed that the technology of media would stagnate in the 1970s even as the technology of space living soared into the twenty-first century). At the same time, Heppenheimer was content to leave waste management to the "Zimmerman process," where wastewater was heated under pressure and returned to galactic farms and space apartments. He also had confidence that the management of waste in space colonies would be modeled after the process aboard submarines.[18]

Heppenheimer was obviously unaware of the fact that the designers of the Algatron, sanitary engineers William Oswald and Clarence Golueke, had already dismissed this idea in 1961 (and others agreed with their assessment). Submarine systems were incompatible in space because submarines and space ships were simply too different in terms of launch-weight and duration of the voyage. Furthermore, microbiologists recruited to NASA had noticed in a test inside Langley's Integrated Life Support System as early as 1967 that microorganisms entirely displaced normal mixed flora. This was in "marked contrast" to "studies conducted in Antarctic on small groups of men cut off from physical contact with other communities," like submariners.[19]

O'Neill shared the breezy assumption that life support would be easily solved at some point. One hundred years from his writing, that is, by 2081, O'Neill mused, "the development of closed-cycle (greenhouse) agriculture will be the only effective solution to the problems of agricultural pollution and deforestation and also of year-to-year variations of food production." Yet he had no specific ideas about how such agriculture would function or how the tremendous difficulties in making these cycles work would be resolved. Blithely ignoring over a decade of work at NASA, O'Neill simply stated that, obviously, "water once introduced into a space habitat will remain, circling through a closed ecological cycle." The views of fancy visioneers versus of square engineers had a direct encounter on stage in 1976 during a public debate over the nature of future space colonies. O'Neill starred alongside Rusty Schweickart, the former NASA astronaut, who was by then a frequent correspondent with Stewart Brand and visitor on Jacques Cousteau's boat. The pair ended up discussing the practicalities of life in space, including the thorny question of how to manage water, watersheds, and waste in a low- or no-gravity environment. O'Neill had little to say. In contrast, Schweickart regaled Warshall with a detailed description about how to defecate in space—presumably the most direct answer to the problem of waste in a closed system that Warshall ever received.[20]

THE FOUNDATIONS OF CELSS

Unimpressed by the visioneers' stirrings, NASA got on with business. The institutional crisis of the mid-1970s led to a distinct change in direction. Delaying another *Skylab*-style space station and the building of any further Saturn 5 model rockets, NASA turned instead toward a reusable, low-orbit space transportation system (STS), better known as the space shuttle. NASA considered the space shuttle a practical intermediate step between the initial triumphs of the Moon landing and that unknown goal which would come next. The shuttle promised to make entering space a reliable and regular service to deploy satellites and scientific instruments, and there was an expectation that both the number of people and the length of their stays would gradually increase. Yet it remained an unglamorous project in the shadow of Apollo. The U.S. Air Force dismissively regarded it as a space truck, in reaction to how NASA described its function. Seeking to routinize space, NASA embraced the prosaic image of a new generation of astronauts "going to work" via a shuttle. In a more colorful flourish, just before the first shuttle was unveiled in 1976 its name was changed from the stoic *Constitution* to the bold *Enterprise*. The new name resulted from a petition campaign by Trekkies, which was proudly handed to President Gerald Ford. Members of the original *Star Trek* cast were invited to the ceremony and a military band played the theme from the television show.[21]

The shuttle obediently did its work. STS missions hauled whatever NASA required into space after 1982, but both the shuttle's load capacity and life-support system were limited, so it was only suited for travels into low orbit. All of this made people question whether it really was the appropriate first link in a chain that led to lunar bases or missions to Mars. The shuttle was always "in danger of becoming the Edsel of space transportation," a former senior manager put it. An even more damning view came from President Jimmy Carter, who noted in his diary in June 1977 that the space shuttle was a "contrivance to keep NASA alive." Even within NASA, the general feeling was that the shuttle was never quite big enough or could stay in space for long enough to push the post-Apollo frontier very far. As late as 1993, NASA director Michael DeBakey testified before a house committee to the problems imposed by the space shuttle's inability to stay in space for longer periods: "We do not limit medical researchers to only a few hours in the laboratory and expect cures for cancer," he bluntly explained. "We need much longer missions in space—in months to years—to obtain . . . new knowledge and breakthroughs."[22]

Research and development of life-support systems remained a key area at NASA precisely because the shuttle promised that both the number of people and the length of their stays in space would gradually increase. NASA's biological mission in space centered on the CELSS project, which was initiated at NASA's Ames Research Center in Palo Alto in 1978. By the 1970s Ames's location in the Bay Area proved an asset, as it found itself in the immediate proximity of some of the most significant scientific and technological revolutions of the late twentieth century. Ames started to prominently collaborate with the growing computer industry in Silicon Valley on, among other things, aviation safety and robotics.[23] It eventually became a stronghold of astrobiology, including the study of the origin of life and the search for life elsewhere in the universe. The most famous parts of astrobiology became the search for extraterrestrial intelligence and for organic molecules in space. But the new field also turned its eyes to the ground. In order to conceive of other forms of life, astrobiology looked to understand harsh environments on Earth, seeking out the freezing Arctic, dark deep sea, or scorching deserts. Such habitability studies overlapped with NASA's continuous interest in the development of artificial environments and life-support systems.

The CELSS became NASA's major effort in studying the range and limitations of an environment's habitability. The rest of this chapter will describe its development from the late 1970s until the mid-1990s. (Work on closed life-support systems has, of course, continued, aimed particularly at the growing of plants on board the ISS.) It started, like its predecessors, with putting algae systems to work, but, as the project's leader stated in hindsight, "American researchers became disappointed with microalgae and chemosynthetic bacteria during the 1960s and 1970s." Making algae both edible and palatable had remained an outstanding problem, and then there was the perpetual concern about the long-term stability of mass cultures. The CELSS did not fare any better, and starting in the 1980s, NASA turned to higher plants. While academic biology went down to the molecular level, the creators of the CELSS moved from algae cultures to farm crops, with a distinct emphasis on plant physiology, controlled environment biology, horticulture, and agriculture. Just like in the Soviet Union's BIOS facility, "closed system cultivation occupied a central position among the major research directions." The work at CELSS focused on plants that did not require cross-pollination, had a high proportion of edible mass, and offered significant percentages of daily carbohydrate, protein, and fat for astronauts' diets. NASA plant scientist Raymond Wheeler and his colleagues spent decades on searching the most promising crops, including potatoes, wheat, peanuts, sweet potato, radishes, and rice. They also worked toward regenerating vegetable and "human wastes" that could serve as a "source of water and nutrients for plants."[24] But let us start with its very beginnings.

LAUNCHING CELSS WITH A NEW SPACE ECOSYNTHESIS

NASA's CELSS program launched in 1978. It was codirected by Robert D. MacElroy and Maurice Averner, who brought complementary expertise and experience to the project: MacElroy, a biochemist, had joined NASA in 1970 as part of the Biological Adaptation Branch at Ames; Averner, a molecular biologist, was newly recruited from the University of Colorado School of Medicine. The two new directors offered a major programmatic statement for the years ahead entitled "Space Ecosynthesis: An Approach to the Design of Closed Ecosystems for Use in Space." Above all, the authors made it clear that settlements of the size that O'Neill had predicted overestimated the capabilities of the space shuttle and tended to downplay the reality of living in space.[25] Work in this department was dearly needed. Although the creation of appropriate ecosystems was by then a well-established research theme at NASA, the new directors found the results so far hugely unsatisfactory.

MacElroy and Averner envisioned an initial space station of some 8,000 cubic meters. It would enclose a human living area of about 1,200 square meters and a crop area of 6,400 square meters, sufficient to host the expected crew of seven. A rotating structure that generated 0.5 g would supply a partial gravity environment, while the energy would come from a "large assemblage of solar panels." This was much larger than any spacecraft or artificial environment that had been created so far—by necessity, as MacElroy and Averner explained. The space vehicles deployed by the Americans and the Soviets were much too small to realistically incorporate a functioning life-support system. To be sufficiently stable and reliable, MacElroy and Averner argued, a closed environment required a substantial "buffering capacity" that exceeded by far the volumes that were theoretically required. Even then, control would depend on "constant monitoring by sensing devices and periodic analyses."[26] It was not a piece of self-regulatory nature they set out to construe, but a technologically advanced farm that had to be carefully managed.

This was not going to be easy. As MacElroy and Averner acknowledged, "a catalog of the kinds and numbers of organisms, of the geological, meteorological, climatological, and other inanimate parameters of a volume of space is not a sufficient description of an ecosystem, because it does not include the *dynamic behavior* of the components, particularly the living components." MacElroy and Averner had yet to find the best route to attain the required, deep understanding of the many interdependencies. Significantly, however, they did not find the emergent discipline of ecology very helpful in this respect—for two reasons. First, ecologists had distinctly different aims and interests. MacElroy and Averner found that theirs was a "problem removed from the concepts of classical ecology," since they were not interested in the entangled processes of life as such but specifically in supporting human life in space. Humans were the central component in MacElroy's and Averner's scheme, and everything else had to be developed according to a human's needs. Second, MacElroy and Averner noted that ecology had almost exclusively focused on the behavior of ecosystems "at the gross level of large animal and plant interactions." Ecologists had, for example, developed models of "predator-prey relationships" to understand the driving force of "competition for a mutual requirement, such as food." But they had completely neglected "the microbial level," MacElroy and Averner related, regardless of the fact that "microbial metabolism comprises a significant fraction of the total metabolic activity of the earth."[27]

SPACE SHUTTLES AND STATIONS WERE BIOLOGICAL PROBLEMS

MacElroy and Averner traced their space ecosynthesis ideas back to the algae work by James Eley and Jack Myers from 1964 and to the contracted life-support projects led by Boeing and General Dynamics from 1966 (as we described in chapter 1). In fact, in its first years, the CELSS group continued to do experimental work on photosynthesis in algae, while MacElroy and Averner also started to develop mathematical models of algae cultivation. Full-scale technological development beyond these modest endeavors was impossible to achieve in the budget climate of NASA in the Carter era, much to MacElroy and Averner's regret. NASA nonetheless developed a new substantial Biological Systems Research Program as they looked to return people regularly to space.[28]

In January 1979, under the auspices of NASA's Biological Systems Research Program (BSRP), a workshop assembled more than seventy scientists and engineers from a wide range of affiliations in order to discuss the state of the field. Through programs like the BSRP, NASA attracted and contracted its research and development needs. In the context of regular space shuttle flights and the hope of building an American space station, the BSRP workshop aimed to establish guidelines to direct the research and development of CELSS. Held at NASA Ames, the workshop was the first major conference directed toward a future life-support system in nearly a decade.

As the subsequently published proceedings indicate, closed-ecology life support had continued to be a significant topic for both NASA and its industrial contractors. The cast of characters had also remained surprisingly stable. Along with MacElroy and Averner, the group included Algatron co-developer William Oswald and space ecologist Bassett Maguire Jr. The latter was the son of New York Botanical Garden director Bassett Maguire Sr. and had already been advising NASA on space colonies for a long time. Maguire Jr. was a close colleague of Jack Myers at the University of Texas at Austin, and he had replicated the work of microbiologist Clair Folsome by building closed ecospheres from aquatic plants and animals. Like MacElroy and Averner, Maguire explicitly criticized the "often-published picture of the inside of a space colony—complete with rolling hill and babbling brooks, [and] open fields" as "highly unrealistic."[29] Clearly, they all saw CELSS and the BSRP as part of square NASA.

One group that was conspicuously absent from the conference, in line with MacElroy's and Averner's negative assessment, were academic ecologists: not one of them was listed among the many expert invitees. This did not imply, however, that the group found it unnecessary to look into the interrelation of life processes. It was admitted right in the first part of the proceedings that much more biological research was required: "In past experience with physiochemical subsystems," the proceedings explained, referring to the hurried use of physicochemical systems during early Gemini and Apollo missions, "there has been a tendency for hardware and mechanical developments to outpace the basic understanding of the processes involved." The complexity of the spacecraft's environment, thereby, had been neglected. "Significant gaps exist in the theory of control of nonlinear systems," the proceedings noted, and the gathering therefore recommended basic research into individual and integrated biological processes as essential to any future life-support system.[30]

To this purpose, the workshop—and the BSRP at large—advocated building a ground-based life-support system as a first step, to be tested with humans. They conceived of a yearlong experiment with twenty-four occupants. This, they thought, would be a "relatively ambitious biological model for in-depth study of the fundamental dynamics of ecosystems, providing information on ecosystem complexity, structure, and functioning." Beyond that, however, the views of the workshop's five working groups diverged. Some, though unnamed, of the workshop's systems engineering and modeling group were concerned that a ground-based test system could not "address the issue of human behavior under CELSS conditions in space." No one in the waste processing, food production, or ecology systems safety groups chose to mention that drawback. Instead, those groups emphasized suggestions to close the cycles of energy and material. One approach advocated oxidizing all organic wastes to produce CO_2, and then releasing this gas into a growing chamber to be converted by plants via photosynthesis into oxygen. Another was an "integrated algal bacteria system" to process all wastes, or perhaps the system might turn to "using higher plants to grow directly on the soluble waste solution."[31]

Water, in this context, acquired a special role. As one contributor argued, it might serve as a "carrier fluid," both "within the living components" and "externally in the waste processing subsystems." Beyond its biological role, water could transport the nutrients that were used for growing plants, notably potassium, phosphorus, and nitrogen through their appropriate chemical forms.[32] This appears to have been an influential idea. A later CELSS diagram from 1984 (figure 4.1) included a water system that transported waste from each component (plants, animals, humans) into the "waste processor," while "recycled water" redistributed nutrients back. Among other features, it incorporated fishponds into an aquaculture irrigation and fertilizer system for the plants. At least according to this diagram, the waste processor was the unexpected center of the CELSS: the single largest piece of technology through which everything essentially flowed.[33] It is not exactly clear how the Biological Systems Research Program fed into the

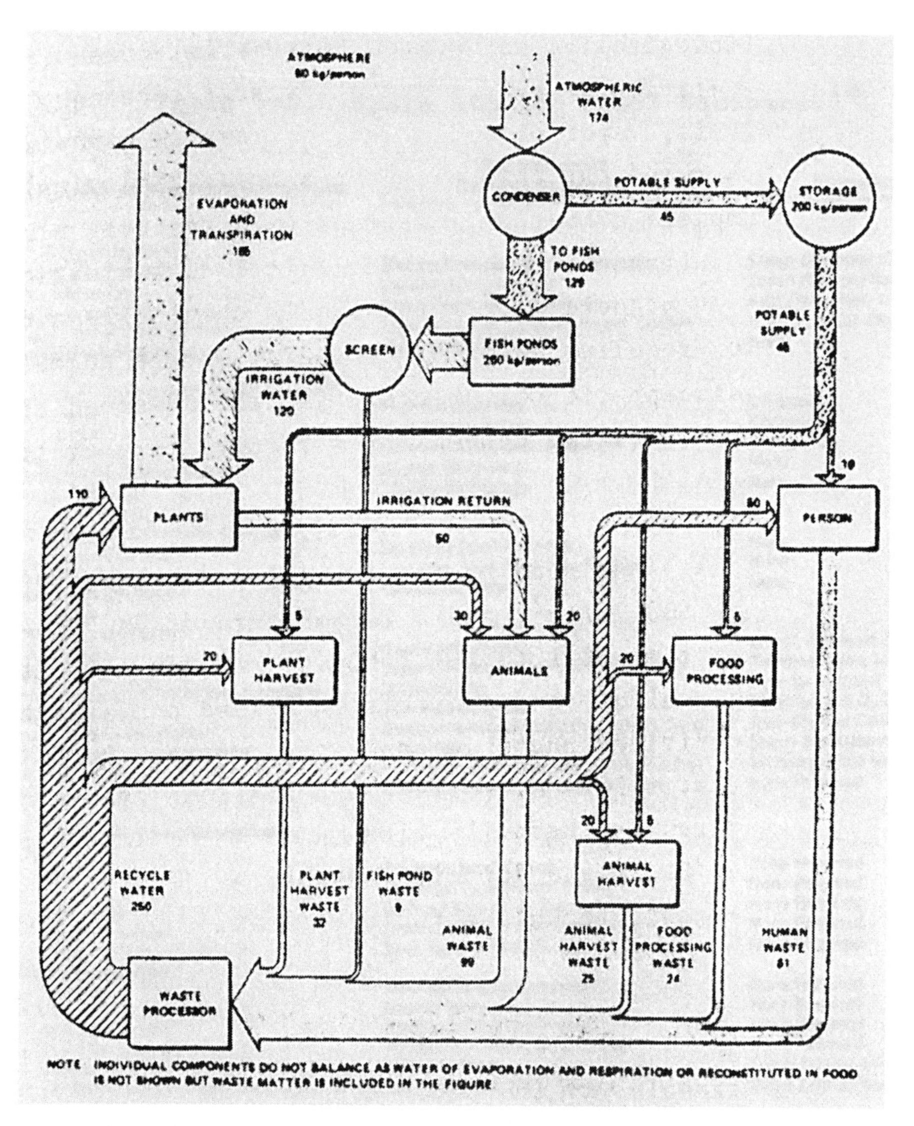

Fig. 4.1. Dual water supply, kilograms per person per day. Shiro Furukawa, "Closed Ecological Life Support System," fig. 5.6, in *Life Sciences Considerations for Long Duration Manned Space Missions*, vol. 1, *Medical Operation*, NASA Contract NAS8-3235, KSC-1, Box 2, R&D Projects—Space Station Program Reports, 1982–84, RG255, NARA (acc. no. 255-09-001), 5–8.

planning of the American space station, but ideas like using water as a transport medium also shaped the next steps toward life in space. Steps that anticipated to be taken on board space station *Freedom*.

FREEDOM: A SPACE STATION FOR THE EIGHTIES

Ronald Reagan, the former governor of California and president of the screen actors' guild, swept to capture the American presidency from an embattled Jimmy Carter in November 1980. The optimism of Reagan sharply contrasted the Carter administration, which was mired in a host of domestic issues including the Iran hostage crisis of 1979. Reagan's administration strongly linked the nation's political identity to the restoration of American strength. The concerns of NASA were little heeded during Reagan's first term in office. The new president addressed conservative chestnuts like cutting taxes and challenging unions, as well as social issues like school prayer. However, NASA administrator James Beggs got Reagan to witness the landing of the space shuttle *Columbia* on July 4, 1982, then convinced him that his political fortunes could be enhanced by a visionary move into space. In the run-up for reelection, Reagan came back to this idea. As part of his State of the Union address of January 25, 1984, he dramatically echoed John F. Kennedy in his calling to Congress and NASA "to develop a permanently manned space station and to do it within a decade."[34]

As we have seen, the idea of a space station had been around since the late 1950s but never became a priority of any administration, not even Kennedy's. Reagan's announcement came on the back of several successful space shuttle launches, but also the knowledge that the Soviet Union had already planted two crewed space stations aloft, *Soyuz* and *Salyut*, from 1969 to 1982 and was already planning an even grander station, *Mir*, for 1986. The United States had to catch up—and, hopefully, surpass. NASA felt again boosted at the highest levels—the new plan reenergized the agency in ways not felt since the heady days of the 1960s. The fact that the station was christened *Freedom* was the cherry on the Republican cake served to the voters that November.

The promise of space station *Freedom* may not be as widely remembered as Reagan's "Morning in America" television spots, but both were powerful tools in Reagan's reelection campaign. NASA did not waste any time. The Ames Research Center staff offered its first "Space Station Program Briefing" in February 1984. Political priorities had not changed over the past decades. In line with earlier rhetoric, the first slide reiterated that the creation of a space station would "ensure civil leadership in space during the 1990s" for America. The main fear was that the USSR would emerge into the third millennium as an even stronger rival to the United States. The slides credited Harrison Schmitt, an Apollo 17 astronaut and senator for New Mexico from 1977 until 1982, with the most pointed statement of this scare. When Schmitt addressed the second Case for Mars conference in 1984, he declared that "an attempt to put Soviet cosmonauts in the vicinity of Mars by October 1992, the 75th anniversary of the Bolshevik Revolution, is not only possible; it is highly probable." Driven once more by a keenly felt Cold War rivalry, the United States had to urgently develop and launch its own space station. Repeating earlier claims of social and economic benefits, NASA expected its goal to again promote the development of "advanced aerospace technologies," "couple maturing international space programs to U.S. space systems," and lead to "increasing prestige abroad and pride at home."[35]

For more than a year before Reagan's 1984 speech, NASA had been gearing up for an announcement about a new U.S. space station. But to be politically useful, it had to demonstrate technological achievements that markedly outperformed the Soviet *Salyut* space stations. To some politicians, as NASA deputy administrator Hans Mark recalled, the clearest demonstration of technological superiority would come from placing a completely automated space station in orbit. At a meeting in early 1982, Mark heard Reagan's science advisor, George Keyworth, argue that "the only reason that the Russians put people in space is that their automation technology was so primitive that they had no choice but to do so." Reagan himself, however, was persuaded to support the space station exactly because it would be crewed—this was sold directly to Reagan as a political benefit. NASA administrator James Begg had emphasized this point in a presentation to the Cabinet Council on December 1, 1983, before the station was included in Reagan's State of the Union speech the following January. This was in line with the general sentiment—everyone at Ames had advocated a crewed station for years as "the next logical step." The new space station would be a scientific laboratory and observatory, a transportation, communications, and data processing node. It was envisioned as an assembly facility to further explore space, and NASA found that it needed a higher capacity electrical system from solar cells (suggesting developmental work on "fuel" or even "nuclear" cells) as well as new data management systems that closely monitored the station.[36]

Because of the BSRP in 1979–1980, Ames was primed to deliver what was suddenly demanded from NASA to gain presidential support, namely the development of a working "environmental control/regenerative life support" system: the logical consequence of planning a crewed station. This suddenly provided a concrete reason for the CELSS program's expansion. Internally, NASA's problem statement from December 1983 said that the space station needed the ability to go "90 days without STS revisit" and "5 days without routine space station ground support." NASA accordingly asked for $150 million to start work on the challenges that the switch from short-range missions to a permanent station would face. A new urgency was put on the development of closed-loop life-support systems because these were the only ones that made economic sense to NASA. In a workshop six months earlier, in July 1983, John Hall and Shelby Pickett from Langley concluded that an open life cycle onboard the space station (i.e., under conditions of resupply from Earth) would cost $560 million; in contrast, the development of a closed cycle was estimated to cost only $84 million. Both figures were calculated over the lifespan of the station. In NASA speak, this resulted in "Δ=476M": that is, NASA would save $476 million if it went for closed systems—and so NASA did.[37]

There was clear excitement in the planning process of December 1983 as NASA worked to secure the presidential endorsement on the back of a realistic conception of the new space station. Any element that would supposedly minimize cost was displayed in bold lettering or in spotlighted text boxes on documents and slides, from the sensible to the farfetched. NASA's Concept Design Group (CDG) gathered all the available ideas and data from throughout NASA, including the BSRP, Ames's Life Sciences directorate, and industrial contractors, to assemble the Space Station Program. The CDG advocated building "common habitat modules" instead of individualized modules to reduce cost but also suggested offering interior design companies the chance to provide the funds for "designing pleasant surroundings" for the space station.[38] There was, however, a specific view of what counted as "pleasant," as Shiro Furukawa, an American medical doctor working for the McDonnell Douglas Technical Services Company, clarified at a meeting with his colleagues from the design department: "The ECLSS [their variant of the CELSS] presents two main technological challenges: to achieve technological readiness for design and to prove that a firm commitment can be made to technology, which has never flown in manned spacecraft. These challenges entail subtle issues of automation, maintainability, commonality, and reliability. They must be resolved in a way that allows the space station crew to treat the life-support system like a utility, requiring only preventative maintenance; it should not occupy a primary position in their day-to-day activities simply because it must sustain their lives."[39]

To Furukawa, any envisaged techno-biological environment, first of all, needed to be robust. The life-support system should function as a "utility," as he put it, which implies a distinct inconspicuousness: reliable at minimal maintenance and hardly noticeable in everyday use. (This, we may speculate, was in particular expected from the waste management system.) The aim to develop a stable, long-lasting environment was one of the few aspects of the space station that everybody agreed on. On almost everything else, opinions differed wildly. As Randolph Ware and Philip Chandler from the U.S. Office of Technology Assessment pointed out around 1983, there was "no such thing as 'The Space Station.'" Rather, a "variety of sets of infrastructure elements" were discussed, "ranging from modest extensions of current capabilities to vastly more sophisticated, extensive, and costly ensembles." Some even envisioned a lunar or Mars jumping-off point. But regardless of scale or style any space station would require "high-reliability, space-maintainable life support systems."[40]

NASA's new impetus in this direction openly took advantage of the substantial Soviet work on closed environments. Part of Ames's Life Sciences directorate contribution to the CDG's planning of the space station was to have studied Soviet literature on space medicine and biology. There was much to learn (as we have seen in chapter 3). Soviet cosmonauts had made it very clear, for example, that privacy was one of the top priorities of personal comfort. NASA's new habitability designs therefore installed curtains into its modules for the space station. Likewise, they learned how to deal with the problem of personal grooming or the astronaut's toilet: NASA collected Soviet techniques for shaving, cleaning teeth and skin, as well as body sponging. Waste on the *Salyut* station, the Americans learned, was collected in their emptied resupply ships, not returned to Earth but allowed to be "destroyed during reentry" (which is still a feature of the ISS cargo resupply system in 2020).[41]

One of NASA's most important industrial contractors, Boeing also assembled an enormous catalogue of the Soviet experience of living in space. Among the top four most "irritating factors" of long-term space living was the absence of water for showering, Boeing noted in their contractor report to NASA in 1983. (This was hardly news to the Americans, who had spent considerable effort to try and perfect their own space shower aboard *Skylab* in the early 1970s, albeit with limited success.) What Boeing found almost as remarkable was that the Soviet space program also had developed a specific fabric for the towels, which absorbed the water both on the skin and on the walls of the shower cubicle. This was an important innovation, as stray water floating around a space capsule could interact with exposed electronics and was considered a high-risk affair. The Soviet space program had also installed a water regeneration system between *Salyut* 6 and 7 (1977 and 1982) that not only recovered the water used by the cosmonauts but also removed "the excess moisture from the equipment." The water recycling system permitted the total weight of the water stored aboard the *Salyut* to be reduced from ten to two tons, Boeing reported. Even better for the cosmonauts, the water regeneration system could produce a limited amount of hot water. Years later, American astronaut William Pogue reflected on Americans' failure to install comparable systems in the earlier *Skylab*. As Pogue pointed out: "If *Skylab* had been designed for repeated visits over several years, then recycling of water would have been practical," but at the time, although moisture was removed from the atmosphere on board *Skylab*, "we didn't use it for anything. It was collected in a wastewater tank."[42]

Like the limited bathing options, food had long aggravated astronauts. Moving beyond fanciful notions of meals in a pill or prepackaged food, NASA redoubled its efforts at perfecting the growth of food in a space environment. One large project subsequently ran between 1984 and 1993, after NASA contracted Ted Tibbitts of the Biotron facility at the University of Wisconsin–Madison to develop "space potatoes." Potatoes had been identified as one of eight potential plants to be incorporated into the artificial environments in space. They were nutritious, had high yield, and also were "palatable and acceptable to most people." Tibbitts's team experimented with more than twenty cultivars of potato, beginning with the white potato (*Solanum tuberosum*), which they found would supply a single person's nutritional requirements in an area of twenty square meters. While they confirmed that certain potato cultivars like the Kennebec required periods of darkness (more than eight hours), many others grew continuously in longer light periods as long as the temperature did not rise above twenty-four degrees Celsius. At the same time, potatoes were also found to grow at the highest rates in elevated carbon dioxide levels. "This crop has distinct advantages that encourage its inclusion in space life support systems to provide food, supply oxygen, purify water, and removal of excess carbon dioxide," Tibbitts reported.[43] NASA finally concluded that through these experiments enough was known about potatoes to use them effectively in space.

Showering and food, of course, were just two elements of the ensemble. Another was clothing. During the CDG planning process in 1983, NASA entertained several ideas for clothing fibers that could be broken down and reused, as had Soviet researchers (apparently to little effect: on the ISS, clothing is currently worn for several days and then placed in the waste area to be burned up in the atmosphere inside the resupply vehicle). Still another was nutrition. The goal of nutrition was to "provide maximum flexibility of resource utilization." This implied that the food would be engineered to "meet the nutritional [and] aesthetic requirements," while it was "prepared from available components (plant, animal, microbial, plant culture, or synthetic)." Food production emphasized "systems closure, component interfaces, waste regeneration/processing, and contaminants." In parallel, waste processing offered two outcomes, namely a "plant nutrient solution" to be used by plants or alternatively some "integrated approach which yields food products without producing a nutrient solution as an interim step."[44]

One major concern during the planning of the space station was how knowledge would be compartmentalized around objects like food or waste. Lists of seemingly discrete features became understood as deceptive: food and waste were not really separate but parts of a continuous process. Building on the conceptual work from the late 1970s, Furukawa explained the real, bewilderingly complex task that lay ahead: "The integration of the thermal, physical, functional, control, and monitoring of these 50 or more diverse functions included in a total ecosystem will be a most formidable task, if not an impossible one. [Worse still,] the output or product of one function must be compatible with the next process, even with greatly varying loads and conditions. . . . History has taught that subsystem comparisons can be greatly misleading (individually tested subsystem data are not usable in a total system) without full consideration of the system level impact."[45]

Of course, there was yet another component, which Furukawa did not mention but which made things far more complicated still: the human within the system. Human behavior in a closed artificial environment remained a great unknown. It was difficult enough to forecast for one person, but when several people were enclosed, as the Soviets had done and the Americans were planning to do, the range of possible outcomes was frightening. As imperfect as such test runs might be on the ground, before NASA could build *Freedom* and send it up into space, its crews needed some controlled experimentation.

A CLOSED PSYCHOLOGICAL AND PHYSIOLOGICAL ENVIRONMENT

Like their Soviet colleagues, NASA had engaged since the late 1960s in elaborate psychological testing of people in extreme isolation or confined environments.[46] Nuclear submarines, and deep-sea and Antarctic research stations became exotic venues for psychologists to study group dynamics, interpersonal relationships, and individual mental health.[47] There was, for example, a series of "*Skylab* isolation" experiments running for 105 days in 1973 and repeated a decade later. One of the participants, Bruce Pittman, reported in a review from 1985 that in a closed environment the "nuances become very important"; in fact, "a squeaky chair is something you'll kill over."[48]

Since then, NASA learned a lot from the widely published Soviet experiences. In April 1974 the main Soviet newspaper *Pravda* reported that two physicians, Mikhail Novikov and Yuri Senkevich, had set out to answer the question "who should be entrusted with an interplanetary spacecraft?" Within weeks their article found its way to NASA's Ames Research Center Life Sciences Directorate. Its timing was provocative, as American astronauts had just departed *Skylab* in February 1974 for the last time, while the second Soviet *Salyut* station had performed well throughout 1973. The two physicians, Novikov and Senkevich, had first joined Arctic and Antarctic expeditions and thereafter accompanied two voyages across the Atlantic Ocean on Thor Heyerdahl's reed raft of 1970, the *Ra II*. The latter, a headline-making endeavor, was a sequel to Heyerdahl's famed

Kon-Tiki raft expedition on the Pacific Ocean in 1947. Novikov and Senkevich used these journeys to observe close-knit, highly driven, and isolated groups of people in unforgiving environments. They speculated on what would "ensure the psychological compatibility of an international crew on interplanetary flights." They concluded that in addition to the physical and technical training, any space flight crew would also have to be prepared to "associate with each other" via appropriate training regimes.[49]

In a different vein, a monthlong experiment in the mid-1990s concluded that to ensure psychological health, crewmembers of long-duration flights had to "have a sense of humor." Self-humor was required in particular: the ability to laugh oneself out of the omnipresent sources of embarrassment, not least in dealing with bodily waste. With otherwise prudish views toward sanitation in space, writers James and Alcestis Oberg noted that throughout the entire lifespan of America's shuttle fleet, everyone at NASA knew that "the shuttle toilets have been the source of vast amusement."[50] Far from the dependable and invisible system demanded by NASA (and the suburban public), astronauts have fought with the facilities, cursed at them, and in the best of cases laughed as they found themselves caught in icky situations involving their waste.

However, as everybody knew, most of the time living in closed environments was not amusing. In August 1983 Jesco von Puttkamer, a program manager from NASA's Office of Space Flight, gave a presentation at a habitability workshop. He made generous use of the diary of cosmonaut Valentin Lebedev's 211-day mission in space—a "*real* long-duration" mission, as von Puttkamer called it—and the resulting paper was full of gloom. Von Puttkamer quoted Lebedev's complaints about the strain of maintaining interpersonal relations, the quality of the food, and the reliance on wet wipes for washing and teeth cleaning. The only positive aspect that von Puttkamer gathered from the diary was Lebedev's opportunity to look through the window at Earth and the stars. Von Puttkamer even presented his audience with a look inside: roughly drawn diagrams that were taken from a "sketch of the Salyut Space Station Interior." They were gathered, as von Puttkamer explained, at a Soviet Industry Fair in Düsseldorf, Germany, in 1982. According to von Puttkamer, the Soviets had worked on the need to make "space a pleasant place to be in." The interior, therefore, was carefully designed, although the result seems a little dubious in retrospect: floors were green-brown, walls were gray, docking ports were painted light blue. The toilet, in contrast, was a bright color (see figure 4.2).[51] Presumably this 1980s color scheme was supposed to have an uplifting effect on the crew.

What is fascinating about the process of defining the problem of living in space at NASA in the 1980s was how wide ranging it NASA's search became.

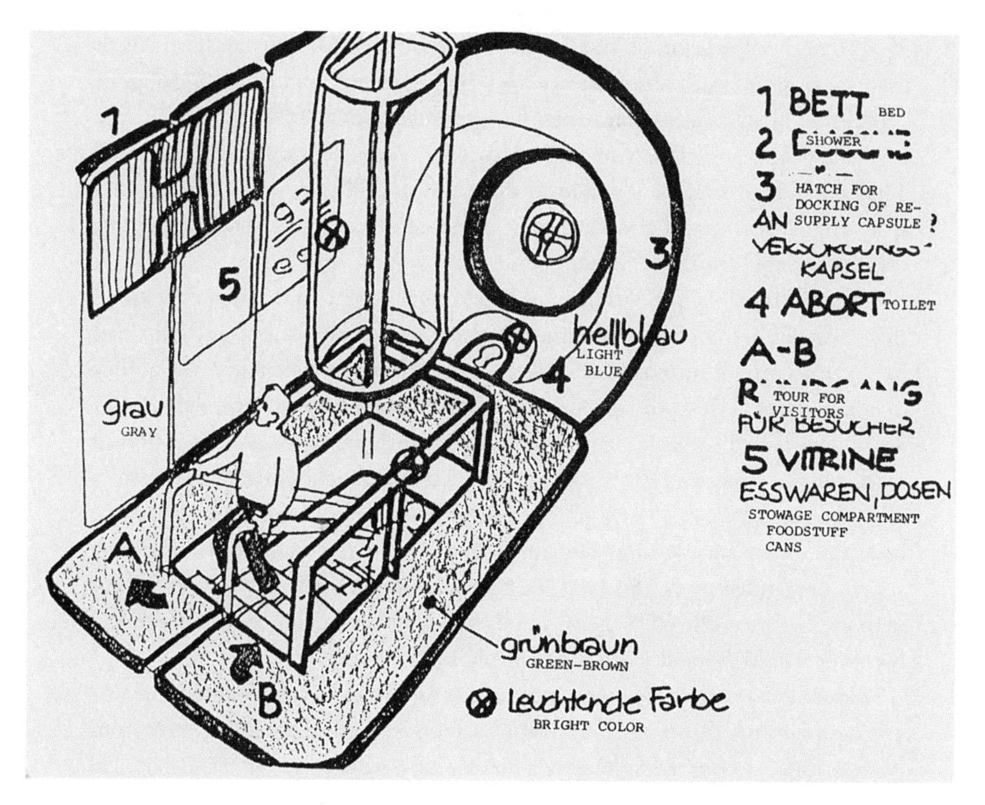

FIG. 4.2. Schematic interior of the Soviet station. From Jesco von Puttkamer, "Human Role in Space," NASA Space Station Concept Development Group, Habitability Workshop, August 30–31, 1983, Volume 6, "NASA Space Station Task Force Concept Development Group," Box 4, R&D Projects—Space Station Program Reports, 1982–84, RG 255, NARA (acc. no. NRHS255-09-001).

In addition to food and waste, having a sense of humor and a pleasant paint scheme, another critical issue was the minimum size of any functioning closed artificial environment. The physiological and ecological aspects were explicitly addressed in the CELSS. However, another problem of long-term space travel was the minimum size of an environment that could preserve a person's mental health. In 1982 Trieve Tanner, chief of NASA's Human Factors Office, reached out to the legal offices at the University of California to ask about the standard square-footage used to house prisoners. The university's law clerk, Sharon Matsumura, replied that rooms of 60–80 square feet with 8-foot ceilings were recommended for no more than 10 hours per day per prisoner, with a large exercise room of 6,000 square feet and 22-foot ceilings plus 35 square feet of

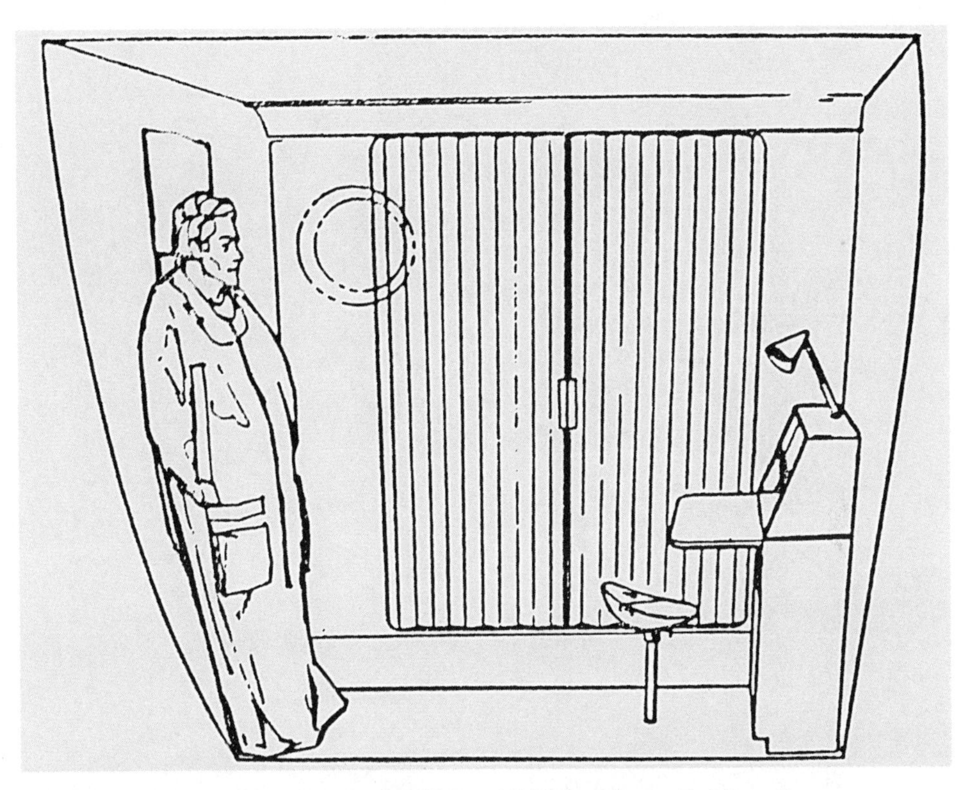

FIG. 4.3. "The Space Station Is Not A Cell." From NASA Space Station Task Force Concept Development Group, "Workshop Briefing Charts," vol. 6, December 5–9, 1983, Box 4, R&D Projects—Space Station Program Reports, 1982–84, RG255, NARA.

leisure room per inmate. One of the startling findings Matsumura reported was a 1982 case from a Louisiana prison that had provided extremely limited space for its inmates—no more than an "average of 15 to 20 square feet" per person. The effects were apparently dramatic: "these conditions increase homosexual activity and encourage aggressive and psychotic or suicidal behavior since there is no territorial space allotted to an inmate."[52]

The psychological effect of the interior became a subject of wide debate. In 1983 Brian O'Leary, a representative from a California aerospace company, Science Applications, gave a presentation to the NASA design group on the topic "What the Space Station Isn't" (see figure 4.3). If NASA wanted heterogeneous crews but also creative scientific research, O'Leary explained, it could not build a station that was a "1960s vintage laboratory," in the tradition of *Skylab* or *Salyut*.[53] An American space station would have plants growing among the stylized furniture

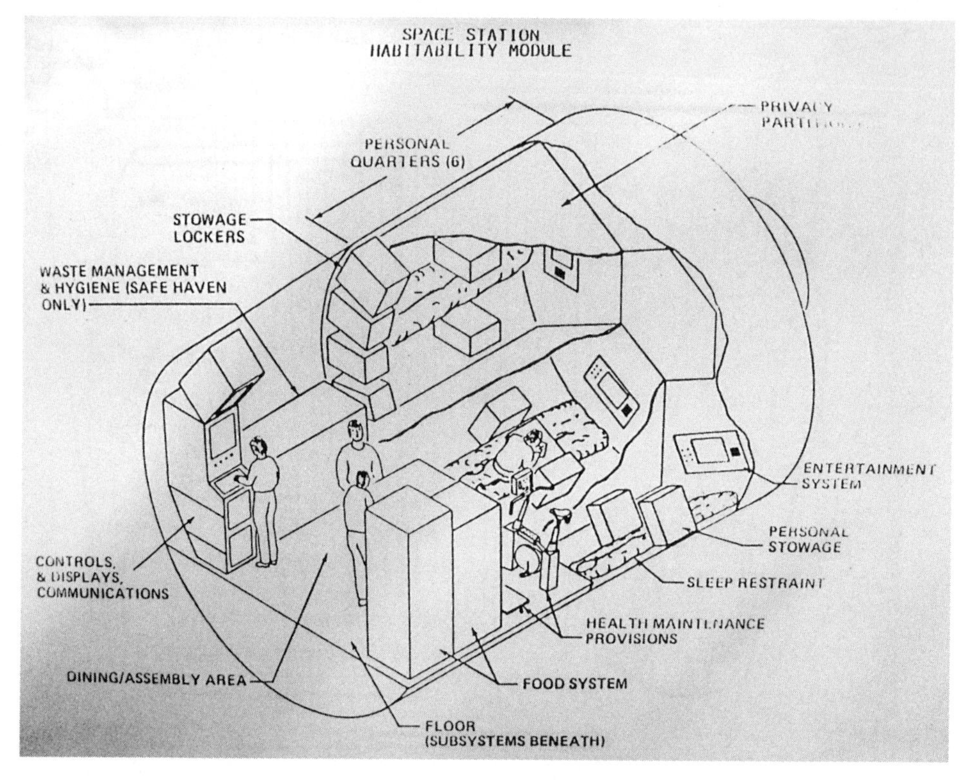

Fig 4.4. "Space Station Habitability Module." From NASA Space Station Task Force Concept Development Group, "Workshop Briefing Charts," vol. 6, December 5–9, 1983, Box 4, R&D Projects—Space Station Program Reports, 1982–84, RG255, NARA.

and trimmings of individual rooms, as O'Leary demonstrated on an accompanying image that looked suspiciously like any science-fiction movie set from the era.

Lockheed's presentation to the CDG by H. T. Fisher offered a similar vision. They also showed off their modular design that stressed private "staterooms" and greater volume for crew habitat to create a more "hospitable environment" (figure 4.4). "Hospitable," Fisher hammered his message home, clearly meant a focus on the details, including "privacy & isolation," "smell & odor," "interior materials," "group behavior," "waste handling," and "hygiene stations." His colleagues from Boeing, Keith Miller and B. J. Bluth, concurred during another session of the CDG's meetings that the demands of the crew had to be balanced against other aspects of the system. It was too easy, Miller and Bluth noted, to simply dismiss the complaints of alleged "prima-donna" crews in favor of the technical demands of the mission.[54] Highly trained and even-tempered astronauts could suffer in cramped conditions for some time—indeed, a remarkably long time, as Lebedev's experience demonstrated. For a space station to be inhabited permanently, however, sacrificing physiological and psychological comfort to a timetable was no longer viable.

POTATOES, LOTS OF ROOM, AND A WASTE PROCESSOR

Despite all this careful preparation, still by the end of the 1980s, no "American" space station had appeared. Once again, a change in political interest undercut the technical expertise NASA had assembled to take humanity into space. Neither NASA, nor major aerospace corporations or public interest groups such as the Planetary Society or the American Astronautical Society could leverage a space station from the different administrations in office. And then, in 1991, the Cold War ended with a whimper instead of a bang. The United States suddenly found itself the sole remaining "superpower"—which left NASA with hardly any institutional strength to gain political and financial support for its ambitious plans. The frustration over the perpetual delay is no better expressed than in Ben Bova's 1992 novel *Mars*, where a "practically sneering" character says, "you Americans waited twenty-five years before you put up a permanent space station."[55]

The envisaged American space station became in the late 1990s what is now the ISS, a joint project with many other countries, including Russia. The first component of the ISS was launched in 1998, and since then it has been the most continuously occupied station of the space age. Notably, the ISS is an international mission, supported financially also by the Europeans and, an especially sore point for NASA, resupplied by those workhorse Russian rockets that launch reliably every few months, as they have done now for more than half a century. The associated experimental work on bioregenerative life-support system also became the subject of international cooperation. By the 1990s NASA worked closely with the European MELiSSA project, the Japanese Institute for Environment Sciences Closed Ecology Experiment, the Ecotron facility in England, and the remaining Russian group at Krasnoyarsk. NASA itself contributed experience from a renewed CELSS test facility at the Johnson Space Center.[56]

The new CELSS facility at the Johnson Space Center grew from several predecessors at the Kennedy Space Center in Florida. The first laboratory originally was part of the so-called Breadboard Project—a small biomass production chamber that was built in 1985 and offered twenty square meters of growing space on closely monitored film hydroponics. Like many studies in crop growth during the 1980s and 1990s, the project took advantage of the fact that in growth chambers the conditions could be freely determined, including CO_2 concentrations, photoperiods, light intensities, or even spectral shapes, to test the results. But the chamber of the Breadboard Project also included facilities for food preparation, biomass processing, and "resource recovery" modules. Between 1986 and 1993, the scientists first tested crops like wheat, potato, and sweet potato individually, before they proceeded with combinations of crops interacting with both microbes and humans. The rationale was always the same: as Breadboard scientist William Knott explained to a reporter from the Space Society, "if you are going to live free of Earth and be more than just a visitor in space, you've got to develop a system that will make you independent."[57] (The reporter, completing knowledge about the closed loop of facilities themselves, likened NASA's new CELSS to the Biosphere 2, which was already taking shape in Arizona and is the subject of the next chapter.)

After the Breadboard Project, NASA transferred the study of plant growth and human interaction from the Kennedy to the Johnson Space Center in Houston. A new facility was built and a key set of environmental engineering and life science experiments were started. It was called the Lunar-Mars Life Support Test Project. The growth chambers measured 180 square meters of growing area, separate from but could be connected to a closed living habitat. The first test person was a young man named Nigel Packham, who in 1996 spent fifteen days inside breathing the air produced by "22,000 wheat plants growing in an adjacent chamber."[58] He had a bed and small desk, and, according to a photograph from NASA scientist Dan Barta, also a Walkman for company (see figure 4.6). In view of nearly forty years of work to engineer space habitats, the design of this first testing series appears underwhelming. It was a basic sealed habitation chamber, where one occupant exchanged oxygen and carbon dioxide with a single species of plant for a single lunar day. But unlike the earlier gas exchange trials with algae, Packham breathed oxygen from wheat, which could also realistically be used as a food source. Water was condensed out of the enclosed atmosphere and fed into the hydroponic nutrient liquid that in turn watered the plants, but the testing focused on several alternative control systems rather than the air exchange process itself. The photosynthetic rate was controlled via light, physicochemical methods, or by altering the level of carbon dioxide in the chamber. NASA reported that the test series successfully demonstrated that, in fact, higher plants could be used for the purpose for respiration, as the resulting system was sufficiently robust.[59]

The more extensive research and development work began on July 12, 1996, with what they called the first half of phase 2, a monthlong experiment with four occupants. Air was recycled via a physicochemical (not biological)

Fig. 4.5. Nigel Packham growing wheat and wheat generating oxygen. Available at images-assets.nasa.gov/image/S95-15597/S95-15597-large.jpg.

subsystem as was the water from the shower, washbasin, and laundry. Wastewater, both urine and humidity, was also returned to the crew as drinking water. Food, however, came from outside the system with a daily drop through an airlock. Solid waste, in turn, was exported via a "fecal transfer" line running from the toilet to the outside. Over the thirty days of the experiment, NASA reported, its vapor compression distillation system reprocessed 182 kilograms of urine and flush water, as well as just over 3,000 kilograms of shower, wash, and laundry water, as one of the major outcomes of the trial. NASA was extremely proud of this success. In the published assessment of the trials in 1996, NASA said openly that the LMLSTP facility had "recycled water for potable use" for the first time ever.[60] The importance of this success was immediately clear: given that the launch of each kilogram into space cost around $10,000, the recycling of 3,000 kilograms of water would save upwards of $30 million every month after the first thirty days. However, the difficulty of this challenge was equally clear in light of the fact that the vapor compression distillation system broke down at day 27, three days before the end of the experiment.

Six months later, with the systems reconfigured and expanded still further, another four terrestrial astronauts stayed inside the LMLSTP for sixty days. The second half of phase 2, according to NASA's description, was the first "functionally similar" trial to conditions onboard the ISS. The water processing had been fixed and, so it was hoped, rendered more reliable. The major difference was that the experiment replicated the rapid pattern of day and night at the ISS. Orbiting every 90 minutes, the crew on the ISS experience cycles of 53 minutes daylight versus 37 minutes of night—an unusual rhythm for Earth-bound humans and plants.[61] As we saw in chapter 3, Soviet life scientists working in the BIOS-2 in Krasnoyarsk ran a yearlong experiment on lunar time, namely fifteen days of sunlight followed by fifteen days of darkness. In both experiments, though separated by nearly twenty years, scientists sought to explore how growth and development adapted to radically different environments.

The grand third phase of the LMLSTP assembled all components of the previous trials into a single, connected unit. For ninety-one days in 1997, the four-person crew of three men and one woman existed on fully recycled air and water and ate food produced in plant growth chambers (accounting for about half of their calories). For ninety-one days, the total array of biological and technical elements came together at NASA for the first time. Inside a closed environment, humans coexisted with wheat, organic waste became organic fertilizer, cycles of oxygen and carbon dioxide were stabilized, physicochemical and biological systems operated together. Inside that closed environment water flowed always, through humans and plants, through pipes and condensers, through the air and the soil. In a first for the Americans, the crew fertilized their crops with their own feces, processed through a new Solid Waste Incineration System. Unlike the earlier attempts at processing fecal waste through algae or bacteria, the solid matter was burned to render it mostly into carbon dioxide, which was reintroduced into the growth chambers, and some byproducts. As NASA noted, only in these third-phase trials were "human test subjects [integrated] with biological and P/C [physicochemical] life support systems."[62] The use of organic waste as nutrients for crops was hardly a new concept for the space program's scientists and engineers, but putting this idea into practice was far from a matter of course among crewmembers.

Like their Soviet forebears, the Americans harvested wheat and ground flour to bake bread inside the facility. The crewmembers adopted a single washer-dryer for clothes and linens and used the remaining space to build their GARDEN, that is, the Growth Apparatus for the Regenerative Development of Edible Nourishment. One of the crops grown there was lettuce, yielding about four heads every eleven days—not a lot, but a welcome addition to the otherwise frozen food available to the crew. The human habitat was physically separated from the plant growth chambers, so that carbon dioxide was collected and concentrated in the human section before being transferred to the plants. In turn, the oxygen generated by the plants was also collected before being fed into the human habitat. However, the carbon dioxide–oxygen cycle only supported one of the four crewmembers, so that physicochemical systems processed breathable air for the other three. At the same time, the water recovery system used "bioreactor (aerobic digesters)" as the primary treatment for all wastewater, including urine. The biological agents that oxidized the nitrogen were tied to physicochemical subsystems which removed inorganic salts before the water was "reused by the crew." Across the ninety-one-day trial, NASA noted, an initial eight-day water allocation cycled ten times through the crew and their life-support systems.[63]

Donald Henninger of the Johnson Space Center summarized NASA's assessment of this trial: "integrated testing of life support technologies with humans allows for evaluations of their efficacy to provide sustenance to humans. Such tests allow for demonstration of technology-to-technology interface compatibility and end-to-end functionality and operability of life support hardware and software."[64] In other words, if we translate Henninger's technocratic Newspeak: it was a smashing triumph. After having struggled for decades to develop bioregenerative systems that returned waste as nutrients, NASA now had successfully tested just such a system.

CONCLUSION

The successfully tested system never made its way into space. Forty years after closed bioregenerative systems were first recognized as a necessity for a space station, the ISS still uses only a fraction of the full potential to recycle water, air, food, and waste. The ISS remains firmly tethered to resupply from Earth, though its crews over the last twenty years have continued to unravel what it is like to live in space. Indeed, from NASA's perspective the ISS is not so much the first

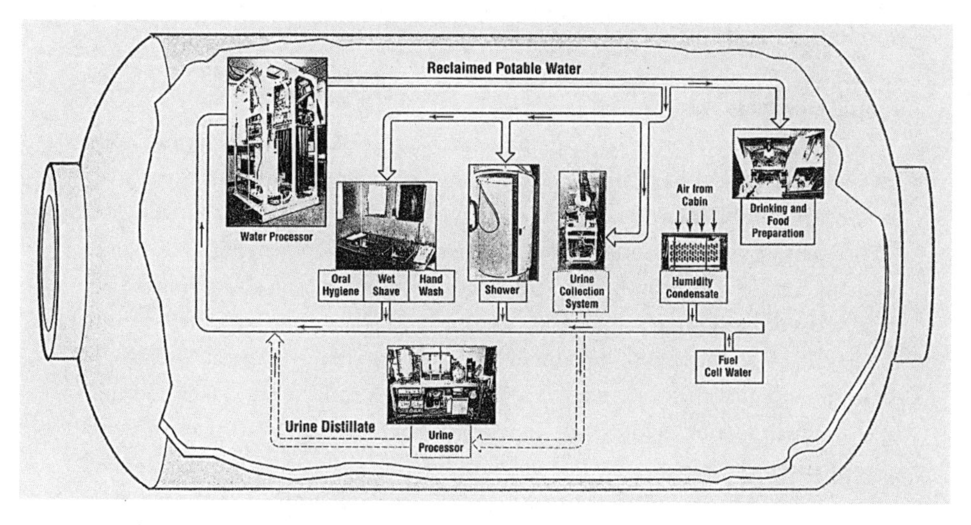

Fig. 4.6. ISS Water Recovery and Management. Available at images-assets.nasa.gov/image/0102168/0102168-orig.jpg.

true space station but rather the next in a series of test modules working always toward a real space station and, finally, Mars missions. We have seen several of these test modules in this chapter; another of them was the Biosphere 2 facility in Arizona, as we see in the next chapter.

Of course, the ISS has integrated some of the LMLSTP components (figure 4.6). Most of the water is processed and returned to the crew, including urine. The solid-waste incinerator module, however, proved one step too far. Currently human excrement and worn clothing is still loaded into the empty resupply vehicle to be burned up in the atmosphere. Still, life in space beyond the mere maintenance has made great strides and garnered a lot of attention: Scott Kelly was surprised, he said, by the public reaction to eating lettuce grown on the ISS (and watered with their own wastewater)—stories about lettuce outshone the spacewalks of his comrades. Having dealt with lettuce, American astronauts toyed with the idea of turning to other plant species: Kelly grew zinnias during his time on the ISS and suggested, "maybe we can grow tomatoes." His Russian crewmates immediately responded, "Growing tomatoes is a waste." They knew that "if you want to grow something you can eat, you should grow potatoes."[65] Mark Watney, forgotten on Mars, would no doubt agree.

5 ESCAPING EARTH IN BIOSPHERE 2

> The ecologists and engineers couldn't insult each other if they tried. The languages they spoke, their methods, and their thought processes were too different.
>
> **—Mark Nelson, 2018**

IN 1984 ENGINEER AND ECO-ACTIVIST JOHN ALLEN AND A NUMBER OF FRIENDS and colleagues launched the Biosphere 2 project with the announcement: "we are going into the space race." Although the story of Biosphere 2 has been told many times, the deeper context of this remark has never been explored. The history of the Biosphere 2 has instead been situated almost exclusively into the context of the then-emergent academic discipline of ecology. And while there certainly were connections, that strong emphasis is in need of rebalance. The Biosphere 2 facility was, in fact, the apex of the long story we trace in this book: the building of artificial environments to support life in space. As previous chapters show, at the time of the facility's creation, biologists and engineers both in the United States and the Soviet Union had long been working on similar projects. Allen and the early biospherians, that group that gravitated to the Biosphere 2, benefited enormously from both nations' first efforts in closed-environment life support as they built the largest closed-environment life-support facility ever constructed.[1]

THE CREATION OF BIOSPHERE 2

John Allen is certainly among the most colorful protagonists in the long history of building artificial environments. Having worked in the steel industry developing new metal alloys, Allen fled corporate America in 1963. He went

FIG. 5.1. Biosphere 2 complex, Arizona. Author's photo.

on a two-year journey around the globe, joined a theater company, and started to write poetry, science fiction, and several autobiographies. In 1969 Allen cofounded Synergia Ranch in New Mexico, which developed and supported projects in sustainable architecture, agriculture, and lifestyle. Allen was joined there by Mark Nelson, a former pre-med and philosophy student from New York, and attracted the attention of the multimillionaire and environmentalist Ed Bass. Bass and Nelson were twenty years Allen's junior but equally concerned about humanity's future—and equally attracted to bold, large-scale projects. Together, Allen, Nelson, and Bass worked doggedly at projects around the world aimed at reformulating humanity's relationship with its environment. In 1973 they cofounded the Institute of Ecotechnics, which was designed as an ecological think tank, devoted, as the founders stated, to "the ecology of technics, and the technics of ecology."[2] Among the institute's most visible efforts was the sailing ship *Heraclitus*, used both as a research vessel and for the drama group Theater of All Possibilities. Starting from the institute, Allen, Bass, and Nelson created an

FIG. 5.2. Interior aquatic/temperate habitat. Author's photo.

organization called Space Biospheres Ventures to build the Biosphere 2 between 1987 and 1991: a huge, fully controlled life-support ecosystem to live in.

The story of Biosphere 2's creation exists primarily in the participant's accounts. They agree that the Biosphere 2 was first conceived in France, at the Cosmos Conference of 1983, another one in the series of conferences that the Institute of Ecotechnics organized each year. By 1983 these events had developed into a who's-who gala of ecological thinking, starring among others biologist Lynn Margulis, famous for many things, including her contributions to the endosymbiotic theory and her support of the controversial Gaia hypothesis. One of the speakers that year was the London-based "green" architect Phil Hawes, who outlined in his paper the components of a sphere that was traveling through space and contained all of life's processes. That very afternoon, the story goes, Allen and Nelson sat down with Hawes and systems engineer Bill Dempster and worked feverishly to calculate the approximate plant density needed to generate enough oxygen for a single person in such an intergalactic sphere. Later that weekend, Allen, Nelson, Hawes, and Dempster began discussing biomes, taking the original concept of a habitable sphere and breaking it into a number of separate ecological systems.[3] At the same conference, Margulis also introduced Allen and Nelson to biologist Clair Folsome. Folsome, the director of the Exobiology Laboratory at the University of Hawaii, had sealed ocean water in small glass spheres, and these small ecosystems had forged new balances of gases and microbes, which proved surprisingly stable. In fact, they sat for years on Folsome's windowsill. To Folsome, the material closure was critical to determine what regulated such systems of varying diversity and stability. He already saw the "far-reaching implications" for "planetary colonization," if a closed and stable ecosystem could be built.[4] As the story has been retold by Allen and Nelson, one thought apparently generated the next, and the idea of actually building a Biosphere 2 started to crystallize (Biosphere 1 was, of course, our planet Earth).

Biosphere 2 was born in the desert of Arizona outside Tucson, and it grew to become a massive 3.14-acre ecological laboratory. Some 185,000 cubic meters of air sat under nearly 16,000 square meters of glass, alongside a million gallons of water. The whole facility cost some $155 million. It was divided into five biomes, or ecosystem regions, including desert, rainforest, agriculture, water, and temperate, linked to each other by air and microbe exchange. Solar energy powered the plants, but manmade energy powered the external cooling systems: a vast collection of pumps, fans, and underground pipes. Among its innovations, the facility incorporated a pair of large "lungs," flexible rubber membranes inside an oval space. Heated air (occupying a larger volume) could expand into these spaces. The underlying rationale was that in Arizona the volume of air would change substantially corresponding to the equally great changes in temperature over a day or a year; installing these "lungs" would keep the facility from exploding from the pressure. Dempster, who became one of the chief engineering advisors, explicitly pointed out that these "lungs" could be of direct benefit to the ongoing CELSS program run by NASA (see chapter 4), in view of the fact that temperature changes in space were even more dramatic. As Dempster explained, it was extremely difficult to ensure the integrity of the air supply on long-term space travel, because the general pressure differential as well as changes in pressure from heat or cold tended to produce small leaks. Installing a set of biospheric lungs might help, he thought. Finally, Biosphere 2 had extensive computing systems installed in the Mission Control Center, which constantly monitored all parameters and collected the data. And there was a lot of data—two thousand sensors took 360 readings per hour throughout the complex, including the temperature of the soil.[5]

Biosphere 2 became widely celebrated, during the 1980s and thereafter, for its combination of biological and technological expertise to create an artificial habitat. The Biosphere 2 was a place, two participants' account enthusiastically noted in 1991, where "state-of-the-art technology and ecology go hand in hand."[6] Readers of this book know, however, that, by 1984 endeavors along these lines were already looking back on a thirty-year tradition. Allen and Nelson themselves were closely familiar with this tradition and repeatedly stated that the Biosphere 2 was an Earth-bound analog and experimental trial of a life-support system for space. In accordance with their predecessors in space science and technology, they repeated the well-rehearsed line that the need for such a system derived "from a simple calculation" of the daily requirements of a person in space, which had to be launched from Earth into space (see chapter 1). In a 1992 paper, Nelson, Allen, and Dempster proclaimed the Biosphere 2 a realistic prototype of a lunar base or a mission to Mars; indeed, its creators said that they had quite consciously tried to combine this purpose with the well-established "eco-engineering" expertise from NASA, the European Space Agency, and the Soviet facility BIOS-3.[7]

Besides its physical dimension, the facility was also more complex, as it became the home of thousands of organisms, from bacteria to small primates. Biological experts from agriculture to zoology were involved, and they set out to join forces with engineers and architects. Once again, just as in the early days of NASA and the Soviet BIOS facility, scientific and technological experts came together and discovered how difficult it was to join biology and technology. The control of environmental parameters proved as challenging as the selection of plant and animal species, and the creation of their habitats. Even if the engineers (and evidently some of the ecologists) found it hard to believe, "no ready answer to the exact impact each plant had on the atmosphere" was available. Nelson later described how the biospherians joked "that the ecologists and engineers couldn't insult each other if they tried. The languages they spoke, their methods, and their thought processes were too different."[8] If the biologists sought to create a full desert, rainforest, or savannah by trying to imitate the holistic experience of these ecosystems, the engineers regularly held back in order to control and monitor each and every parameter, from water movements to salt concentrations and the quality and quantity of light. As in all creations of artificial life-support systems in the space age, there was a constant clash between the norms, goals, and expectations of the life scientists and the engineers.

In another repeat of early life-support system development, the Biosphere 2's designers worked on the premise that the elements of the facility were closely interconnected. They assumed that all the buildings, environments, organisms, and technologies all mutually affected each other in feedback loops. To the designers, the entire endeavor was inherently cybernetic in character. "We cannot build Biosphere 2 from fundamental equations upward," atmospheric physicist Carl Hodges said to his fellow designers, because "a true biosphere is a whole interactive 'living entity.' . . . *Every* variable is a *key* variable." Hodges had been in conversation with plant physiologists already in the 1960s, notably with Paul Kramer and Lloyd Evans, who themselves were designing and building new phytotrons, and by the 1980s Hodges insisted on measuring and testing "every variable." The Biosphere 2 designers incorporated organisms explicitly as controlling elements of their feedback systems. Unlike *Skylab* or *Salyut*, the Biosphere 2 relied "on natural helpers" such as ladybugs to ward of insect pests in addition to systems like a "mechanical algae scrubber system" to remove excess nutrients and add oxygen to the Biosphere 2's ocean.[9]

SPACESHIP EARTH, BIOSPHERICS, AND HABITABILITY

Biosphere 2 materialized one of the most powerful metaphors of the new environmentalist movement, namely "Spaceship Earth."[10] The term had been in use earlier but became emblematic in the late 1960s for a number of visionaries to take their message to the public, including widely read books by Paul Ehrlich, Kenneth Boulding, and Buckminster Fuller. It referred to the fact that Earth could be considered a closed system traveling through space with limited resources—and without an emergency exit. As historian Sabine Höhler noted, Spaceship Earth signaled the increasing awareness of the space limitations of the Earth as a life-support system. Additionally, the term *Spaceship Earth* framed "the planet in technoscientific terms and recreated the planet as a new hybrid entity." The term *Spaceship Earth*, in other words, reconsidered the Earth to be a technologically maintained life-support system like those advertised during the space age. To the era, the image of the Earth as a spaceship offered up startling consequences to the future of human life. Architect Buckminster Fuller, for example, claimed that the ability to conceive of Spaceship Earth as a whole structure permitted a technocratic elite to rationalize the planet's entire life-support function. It would fall to the commander of Spaceship Earth, presumably scientists, to decide what was needed to keep the whole system in balance. To economist Kenneth Boulding, Spaceship Earth demanded a new economy that would build on the assumption that the Earth was a closed and finite system, instead of open and infinite. Boulding denounced the wasteful cowboy economy of the past and embrace the frugal spaceman economy of the future. To ecologist Paul Ehrlich, the central theme of the Spaceship Earth metaphor was to question the size and quality of the crew (i.e., how many and which people the Earth could carry and support).[11]

However, the distinct undertones of eugenics that emerged from this discussion of lifeboat economics saw the Spaceship Earth metaphor become unpalatable. It was finally rejected by counterculturalists in favor of local solutions, based on individual decisions or within a smaller community. Ecologist Eugene Odum, for example, replaced the spaceship metaphor in his textbook of 1982 with another reference to space technology, when he implored his readers to consider how "our global life-support system that provides air, water, food, and power is being stressed by pollution, poor management, and population pressure." This to some extent resonated with the Biosphere 2's founding rationale. The creators of Biosphere 2 were not so much interested in a new economy or population politics but did see the opportunity to preserve terrestrial environments. As Nelson noted, the creation of the Biosphere 2 "parallels with humanity's need to create technologies that will permit development without eroding the habitability . . . of our planetary home."[12] But Nelson also circled the justification of the Biosphere 2 back to the ideas of Fuller and Ehrlich, as it would become a facility to test the minimum size of a closed system to keep a defined number of humans, organisms, plants, and environmental parameters alive.

The Biosphere 2 addressed some of the problems that earlier attempts had been unable to solve. In part, the Biosphere 2 operated at a scale unmatched by previous attempts, though at a scale NASA's CELSS and Soviet BIOS designers insisted was necessary. The facility also quantified the effects of a changing climate on various ecosystems and their human occupants and optimized the environment through experimental control of these parameters. The latter was crucial and had been a major obstacle in the past. Unlike closed-environment science, Allen and Nelson maintained, the problem of the science of ecology (especially field ecology) was that it was unable to exert reproducible control over its target systems. They presumably knew about the many earlier ecological experiments that had unsuccessfully worked to trace various components and paths of natural systems.[13] Still, experiments in the field differed from experiments in the laboratory. "Ecologists are in a difficult position," the plant taxonomist Lincoln Constance summarized, "because ecology is really, if you like, field genetics, field ecology, under conditions that make experiment very difficult. It can be descriptive, and often is; but as soon as it becomes experimental, they probably have to bring it into the house and then it becomes something else." The Biosphere 2 project was way out of this predicament because, as Allen and Nelson noted, it was a controlled microcosm in which the immeasurable was rendered measurable. Allen advocated that Biosphere 2 could serve as a control for "Biosphere 1," the

Earth, and this became a key argument to amass further support for the facility.[14] Experimenting on the Earth itself was undesirable, especially if one wanted to push the system to its extremes; in contrast, the environment within Biosphere 2 could be stressed to—and beyond—the breaking point.

According to his own retrospective accounts, Allen thought the Biosphere 2 was not another ecological but the first "biospheric" experiment. Biospherics was, for Allen, a completely new field of study, if not a new discipline. The scientific frontier opened by the construction of closed environments called for the discovery of the "laws of biospherics," Allen said. Biospherics had its conceptual origins, he explained, in the work of the Russian geochemist Vladimir Vernadsky. Allen informed historian Rebecca Reider that he agreed with Soviet life scientists that biospherics was the knowledge needed to live in space.[15]

Allen and Nelson also later acknowledged that their working knowledge stemmed in large parts from the Soviet experts who, he thought, "were leaders in bioregenerative life support."[16] As we saw in chapter 3, Soviet life scientists in the 1960s and 1970s had modeled closed ecosystems in order to understand the mechanism that enabled sustained existence on Earth and had developed and tested closed ecosystems. There was literature available about "manmade ecosystems" from the major figures of the Soviet project including Boris Kovrov, Genry Lisovsky, and Iosef Gitelson. In 1985 Nelson met one-on-one with Oleg Gazenko, long-term director of the Institute of Biomedical Problems and pioneer of space medicine. Subsequently, Allen and Nelson traveled to the Soviet Union to meet Gitelson and the remaining members of the BIOS-3 team at Krasnoyarsk. Gitelson showed them reels of film from the facility, and Allen was able to talk with Yevgeny Shepelev, the first person to live inside a closed environment for a day. As part of the visit, the delegation even observed firsthand the algae-based system, which, as the U.S. team later reported with admiration, had "achieved six-month closure with a dozen food crops supplying half the food and providing nearly all the air and water regeneration for crews of two and three people."[17]

In short, one of the most important sources for the creation of Biosphere 2 was the experience gathered from Soviet life scientists. Allen's high respect of the work of his predecessors in both the Soviet and the American space program is obvious from how he described the project's genealogy. In 1991, shortly before the first Biosphere 2 mission, he cited both the BIOS-3 facility and the CELSS as immediate and congenial forerunners:

> Gitelson's BIOS-3 was large enough to support two or three crew members, and produce about half their food, but suffered a leakage rate of about fifty percent per year. Since it was a test of the microgravity conditions of space

> flight around planet Earth where direct penetration of the Sun's rays would not be possible, its energy for photosynthesis came from high-intensity sodium artificial lights powered from energy from the outside. For the same reason, artificial light had been used in NASA's remarkable CELSS (Controlled Ecological Life Support Systems) program, launched in 1977. Among the researchers involved in CELSS, Frank Salisbury and his team at Utah State University had boosted plant densities for wheat up from the 200–300 plants per square meter usual in field plantings up to 2,000 per square meter and were aiming for up to 10,000 per square meter. Salisbury anticipated that a farm the size of an American football field "would support 100 inhabitants of Lunar City."[18]

Given Allen's acknowledgment of the intellectual and material debts of the Biosphere 2 to Soviet and NASA closed-environment facilities, it is surprising that existing historiography of the project has mostly stressed the ecological roots of the project. One reason may be the rhetoric of the ecologists themselves, who were mostly uninformed (or dismissive) about the work already done within the space programs to build closed bioregenerative environments. Back in 1971 ecologist Dennis Cooke contributed a chapter on the "Ecology of Space Travel" to the third edition of Eugene Odum's famed textbook *Fundamentals of Ecology*. In this chapter, Cooke claimed: "The fact that we are not now able to engineer a completely closed ecosystem that would be reliable for a long existence in space (nor can anyone predict when we will be able to do so because we have not yet given it serious attention) is striking evidence of our ignorance of, contempt for, and lack of interest in the study of vital balances that keep our own biosphere operational."[19]

Cooke and Odum had been a part of NASA's own closed-environment life-support development during the mid-1960s, so for Cooke to claim that closed ecosystems had not received serious attention was both spurious and significant. In part, Cooke and Odum's own studies were called into question both in the 1960s but also later. While they gained a small grant to work with NASA between 1964 and 1967, it was not renewed thereafter, the work itself being practically irrelevant if theoretically informed. In 1986 Clair Folsome criticized earlier "microcosm" studies by Cooke and Odum, who had bounded some small living system, but, Folsome charged, had made "no attempt to isolate them completely from material exchange with the outside world." In a similar case, Cooke had also briefly mentioned algae-based trials. To Cooke "thorough ecological analysis" had demonstrated that a biological system with algae was difficult to control: no human-algae gas exchange could be maintained in equilibrium. Again, this view that became so popular in the ecological literature was later contested by Robert MacElroy, director of the CELSS in the 1980s, and Iosef Gitelson, head of the

BIOS-3 project, in their joint book of 2003. They thought it misrepresented their own earlier achievements. Valid objections to the algae approach existed in the 1970s, they admitted, but subsequent experimental work "verified their reliability" and demonstrated that "stable biological mechanisms act in these systems."[20] But while Cooke's chapter has enjoyed a wide reception among historians of science and technology, the retorts by Folsome and MacElroy and Gitelson have not.

We would argue that the strong focus on ecology has prevented historians to place the Biosphere 2 in the long tradition of research in space exploration and habitation. Höhler situated the Biosphere 2 within a lineage that stretched from Clair Folsome's creation of and Howard Odum's 1971 experiments with ecospheres. Historian Peder Anker draws a line to the Biosphere 2 from the visions of building space colonies in the 1970s, which privileges the ecologists' own version of the story of closed life-support systems. In that version, it had been ecologists who had "suggested" how to build closed environments for the new space program of the 1960s. Like Höhler, Anker stresses Howard and Eugene Odum's conceptualizations, notably their "diagramming," "thought," and "designs," about closed ecosystems on space stations and lunar or Mars bases. For all that, however, Anker admits that "ecologists failed to offer NASA a workable short-term proposal for a cabin-ecology system." As we have seen in previous chapters, this was true, not least because the effort would have been superfluous as it had already been achieved by life scientists and engineers. The "cornerstone" of the Biosphere 2 project, Anker argued, was a 1976 paper published by the project team in the *Bulletin of the Ecological Society of America*, entitled "Ecological Considerations for Space Colonies," because it argued that if a closed stable ecosystem could not be built on Earth it would never be built in space. As we have seen, NASA had already long worked on such systems, and the Soviet Union had fully tested their long-term closed habitat, the existence of which was known to at least one coauthor of the paper, namely former astronaut Russell Schweickart. Schweickart was the one who once had explained to Peter Warshall how to poop in space—a tale famously republished alongside "Ecological Considerations" in Stewart Brand's *Space Colonies* volume—but he was also later in charge of selecting vertebrates for the Biosphere 2.[21]

To those within NASA who were aware of the long history of attempts, the Biosphere 2 was just another stage in the development of life-support environments. A workshop of 1989 assembled many experts from within and without NASA, including Maurice Averner, Wendell Mendell, and Thomas Paine. Paine was full of praise for the project, despite its roots in the American counterculture. He saw in the Biosphere 2 "a shining beacon . . . to an expanding future for humanity" because "closed ecology systems can free us from Malthusian limitations." Biosphere 2 was a public display of a human future in heaven as on Earth—and a very welcome contribution to NASA's own efforts in closed ecosystems. Even in the late 1980s, "the goal of a closed-ecology biosphere" was, in the words of Paine, "the least understood and the most challenging" element of the space age.[22]

MISSION 1

Mission 1 took place between 1991 and 1993. More than four thousand species of plants, animals, and microorganisms were enclosed in the structure of the building. The biospherians, as the first crew became known, would have to find out themselves whether plants from the equator region would be able to cope with the varying lengths of the day in Arizona. They also would have to experiment with rain of differing frequency and intensity to allow for optimal growth conditions in a desert environment.[23]

It was, as Allen's memoirs were titled, a "human experiment." Yevgeny Shepelev, pioneer of closed environments in the Soviet Union, came to peer through the glass of the Biosphere 2 and, in view of the lush scenery, recalled his own small room and single chair with only chlorella algae for company. Eight members made the crew: alongside Jane Poynter and Mark Nelson, they were Abigail Alling, Linda Leigh, Mark Van Thillo, and physician Roy Walford. As Alling and Nelson later said, their excitement to be part of the project stemmed from its general goal: "the understanding of the closed sustainable ecologies needed for human habitation in space." Locked inside with his seven companions, Nelson penned a poem after his enclosure entitled "I Live in a Glass House." Whether it was a "Frankensystem or Alice in Ecoland," he wrote, "the creation lives."[24]

The mission seems to have started with a spirit of high adventure, similar to the early astronauts' experience, which possibly helped them through their suffering—at least during the first months. "The eight of us had risked much to live as if on Mars," Poynter noted. Over the next two years they certainly endured a lot, including starvation, social isolation, and the psychological pressure of being part of a small community that depended upon each other. It is almost ironic that they also had time to complain that there was no toilet paper, a fact that seems to have caused considerable discomfort. The recycling system was unable to handle the amount of paper eight people would generate over time, so they had to use a water squirter instead. Oddly enough, Allen still described the requirements of waste management within a closed ecosystem as being like those "conditions on nuclear submarines," though it had long been shown that waste management in space was in fact very different from the one underwater (see chapter 4). Great care was taken to implement functional material cycles. The Biosphere 2 was equipped with a physicochemical water recycling system that resembled a naval unit for potable water. "All of the water that comes from the human habitat—from toilets, showers, kitchens, laundries—goes into the basement of the agriculture area to our waste recycling systems," Alling and Nelson later reported, and they stressed that this system had to be checked every day. Biosphere 2 also had an elaborate wastewater system for animal and laboratory wastes that combined a septic treatment with biological filtering through marshland. The resultant plant growth was periodically cut back and either fed to the animals or composted.[25]

Inside the Biosphere 2, the inhabitants planted their crops and slaughtered their goats. They also sat at computer terminals in the control room and at benches in the laboratories. Although the facility was publicly celebrated as fully computerized and monitored, the biospherians were wary of surrendering control to machines and systems: "I'm very skeptical about computer systems controlling everything," Van Thillo reportedly said.[26] But on closer inspection, the autonomy and self-regulation of the crew was mostly an illusion. As in earlier constructions, starting with the Algatron, the role of the human inside the Biosphere 2 was more maintenance than innovation.

The production of food became the biospherians' most pressing concern for nearly their entire two-year mission. Yet, the disposal of waste presented was, in fact, an even greater threat, notably waste in the form of excess carbon dioxide. The crew noticed after just a few months that oxygen was constantly disappearing from the system, while the amount of carbon dioxide was increasing. They physically removed carbon dioxide by storing plant material, in addition to the work of the carbon dioxide scrubber, and they took to donning scuba gear to weed the algae that grew prolifically and consumed precious oxygen in the pond. In contrast to earlier conceptions of algae as the miraculous substance that sustained life in space, the biospherians described algae as a "terrible pest" and a "significant weed problem throughout the two-year experiment." Increasingly desperate, they mixed the algae with other fodder as animal feed to dispose of it. They did not, however, attempt what the early Algatron, CELSS, and BIOS-3 scientists had advocated for: consuming the algae as food. This is to some extent surprising. Even in science fiction, such a diet is considered unpalatable but sometimes necessary: as a character from a David Brin story notes, after the supply ship blew he had to "live off algae paste [and] wishful thinking."[27]

A THEATER OF SCIENCE

The first experimental run of the Biosphere 2 was longer than any previous attempt to keep humans and other organisms alive in a closed environment.

Mission 1 sealed eight people inside for two years. But while this could have been celebrated as a pivotal moment of environmental awareness and space age heroism, the Biosphere 2 instead became portrayed as a circus. The media latched onto Bass's sponsorship of the Synergist Ranch group and the selection of the first group of biospherians, who were preparing to be enclosed for two years. Portrayals of the project argued that the enterprise was not only unaccountable and unscientific but essentially a cult.

The charge was first leveled by *Village Voice* journalist Marc Cooper. Cooper criticized the scientists involved in advising the project and concluded, "There can be no question that Biosphere 2 has been conceived in full accordance with the same febrile survivalist notions that powered John Allen's dinner-time harangues on the Synergia Ranch: escape from a dead Earth and colonization of Mars." Cooper questioned the scientific qualifications of the founders and the crew.[28]

Cooper's criticism stung, not least because the Biosphere 2's creators had taken pains to stress the wide-ranging credentials of the "New Biospherians." They were advertised as being "plumbers, electricians, and tailors," "good observers," and sportspersons who would easily learn to scuba dive—alongside traditional qualifications such as doctor, field scientist, and trained mechanic. Ironically, even from this laudatory description, the crewmembers seemed to have been specialized in everything except growing plants for food. Nevertheless, very soon "everyone spent two hours after breakfast working in the agriculture system" each day, often more. The unexpected burden of time and labor at this front irritated some of the crew: Nelson recounted that Walford complained after just two weeks that farming was taking him away from medical research. But the problem went beyond a lack of enthusiasm. As Leigh noted, "We just plain were not good farmers." They had to learn this craft by doing, and they had to be quick because they crucially depended on its success. As Poynter summarized the situation, "Farming is bloody hard work," and it was important too.[29]

Several years later, Jane Poynter still objected to Cooper's labeling the core group of Biosphere 2 "a cult," but her own description certainly draws the image of an unusual endeavor undertaken by an unusual group. Poynter fondly recalled her training aboard the vessel *Heraclitus*, the oceangoing sailing ship that was one of the major projects of the Institute of Ecotechnics. Poynter fell "in love" with "her," the Biosphere 2 building, as Mission 1 approached. Alongside John Allen and five other biospherians, she allegedly "howled at the moon, and all but drew knives across our palms in a Ya-Ya blood pact, swearing allegiance to each other, to Biosphere 2, and to the successful completion of the two-year enclosure." How much of this was real or surreal was, it seems, deliberately left open. The whole group also invoked the performance of the astronauts by acting out a parody of the astronauts from Tom

FIG. 5.3. The geodesic dome of *Silent Running*.

Wolfe's novel (and later film) *The Right Stuff*. Thus, it is not surprising that not every report of the project was favorable. There is, however, little evidence that the endeavor was merely one long commercial as Cooper once alleged: in response to a critique of his articles on the Biosphere 2, Cooper maintained that his problem with the facility was that its creators had more "interest in creating a corporate colony on Mars than they do in creating a more livable Earth."[30]

It probably did not help that images and stereotypes from popular science fiction of the 1970 and 1980s seemed to feature heavily in Biosphere 2. Integrations of fiction into science, Evelyn Fox Keller noted, are "hopeful statements" that express a direction of intentionality; she even claims they "actively contribute to the construction of future scientific reality." In this sense, there had been elements of science fiction in the entire history of producing controlled environments, whether it was an Algatron or a lunar base with space potatoes—they all were invoking images of a future they knew was far ahead but still figured as the target of their work. Biosphere 2 pushed that element into new dimensions with its roots in and its rhetoric of the American counterculture. There was, however, an interesting paradox here. The science of the project, notably the aim to elucidate the dynamics of whole ecosystems, was actually fiction, that is, far too complex to be attained in any foreseeable future. The apparent fiction, on the other hand, the epic two-year mission inside a closed environment, was part of down-to-earth science because it allowed for a test run of very concrete life-support arrangements. Most strikingly, the Biosphere 2's architecture evoked a number of widely known science-fiction films, notably the major design elements of the 1972 film *Silent Running* with its geodesic domes (which, incidentally, were themselves copied directly from the Climatron ecological facility at the Missouri Botanical Garden in Saint Louis). *Silent Running* presented a dystopian vision of humanity's future in space, where biodomes contain the last bits and pieces of Earth's organic matter; eventually, these biodomes (and their plants) are jettisoned into space as waste and detonated (figure 5.3). Director Douglas Trumbull later recalled that the film started as a commentary on the Vietnam War but grew into an even stronger commentary of the environmental problems of the era.[31] All in all, it made a perfect point of reference for the Biosphere 2 project.

Finally, the biospherians also performed a display of science fiction for the assembled news media as they entered the facility in 1991 for Mission 1, when they appeared in specially designed Biosphere 2 jumpsuits for the publicity photos. The jumpsuits (which Alling and Nelson said some people called "Star Trek suits") resembled the uniforms of the Visitors from the television mini-series *V*, which first aired in 1983 and then produced spinoffs in 1984 and 1985.[32] *V* enthralled with a story about an interplanetary ecological crisis. The Visitors came from space after their own planet's ecosystem had collapsed to exploit Earth's water and food; humans, of course, were the food.

UNDER PSYCHOLOGICAL AND PHYSIOLOGICAL PRESSURE

Inside the Biosphere 2, two forces conspired to place unanticipated pressure upon the crew and drove them into two hostile camps. First, the role of the humans inside the facility came into question. To Nelson, "the core issue was whether the primary goal was to work on maintaining as self-sufficient a closure as possible and improving the facility [or alternatively] to lessen workloads by sending in food to give more time for research."[33] Some crewmembers thought the labor of farming was simply superfluous: dull work with poor results. To others, developing appropriate farming techniques became an essential element of the endeavor. Within weeks, the disagreement erupted into an impassioned conflict between two groups of four biospherians each. The resulting psychological tensions have since captured the attention of historians and fiction writers as a forceful lesson in group dynamics, organizational hierarchies, and the place of individuals within large technological projects—notable examples being Kim Stanley Robinson's Mars trilogy or T. C. Boyle's *The Terranauts*.

When the level of oxygen in the facility began to fall, the viewing public saw algae and cockroaches bloom and people go hungry. Over the decade after the mission ended, it was discovered that carbon dioxide had seeped into every nook and cranny of the system. It was even found deep in the concrete walls of the complex, where it had reacted with the calcium hydroxide in the outer layer. The concrete had in fact taken up more carbon dioxide than even the artificial scrubber, which produced hundreds of barrels of stored carbon after only two years. This was an unforeseen complication. The three-man crew of the Soviet BIOS-3 had faced the opposite problem of plants generating excess oxygen as well as lots of inedible plant material—the crew actually burned the dried plant stalks and roots to liberate the trapped carbon.[34] Inside the Biosphere 2, the crew almost suffocated from the opposite effect.

The debate has raged ever since over the sources of carbon dioxide in our atmosphere—the most damaging human causes, possible control of greenhouse gases, and nature's own probable responses. All these issues are now much more prominent than in 1993, when the biospherians emerged from their facility considerably thinner and visibly affected by the experience in body and soul. In retrospect, the main problem was located in the soil. "The designers of Biosphere 2, in their concern that the residents be able to feed themselves, had overloaded it with organic-rich soil," geochemist Wallace Broecker from Columbia University concluded. According to Nelson, the first biospherians had calculated that approximately sixty-seven kilograms of atmospheric carbon dioxide were taken up by "the living biomass" of some seventy tons "in the course of one day." Given the Soviet experience, there was an initial fear that carbon dioxide might become rare inside the Biosphere 2. Instead, the opposite was the case: "bacteria in the soil were consuming oxygen and respiring CO_2. The plants in the greenhouse couldn't photosynthesize fast enough to take up all the CO_2, especially in winter; the concentration had thus risen to eight times the level in the outside atmosphere."[35] In a dark irony, organic farming came close to killing the environmentally conscious biospherians.

Alongside the growing physiological pressure on the Biosphere 2 was a psychological pressure. The founding group imploded in 1994, only part way through Mission 2. For reasons that remain opaque, Ed Bass withdrew his support for Allen's biospherians, and Allen was even legally restrained from entering the property. Soon afterward, the whole complex was sold to Columbia University, then acquired by the University of Arizona. Allen returned with Nelson to the Institute of Ecotechnics, and the Synergia Ranch continued on. Columbia started to distance itself from the (alleged) pseudoscience of space missions to Mars and terms like *synergy* and appointed Barry Osmond to direct projects in

Fig. 5.4. The New Pyramids of Mars. External view of the desert biome, Biosphere 2. Author's photo.

the Biosphere 2 solely geared toward ecology and climate change. Osmond had done much of his early work at the Australian phytotron in Canberra, studying plants under controlled conditions.[36] It was then that Howard Odum began to develop a strong interest in Biosphere 2. Fellow scientists did not in principle criticize Missions 1 and 2, Odum later noted, even if they denied that the biospherian crew did "real science." Odum himself in fact saw Biosphere 2 as proof of the fact that technology could confront and solve human problems. Odum also found a new, more publicly palatable goal for the facility: the study of global environmental change, which, Odum said (as early as 1999!), was an important scientific research program, and so "the Biosphere 2 was reset [via high precision CO_2 monitors] for studies of responses of plants and ecosystems to possible future global environmental change."[37] But Biosphere 2 did not begin as a facility to study environmental change; it only became one after it had revealed humanity's inability to control its environment via technological systems.

The evolution of the Biosphere 2 facility and the attempt at ecological experimentation followed a well-worn path of previous endeavors, which it simultaneously tried to supersede. Its crews labored to map the complex operation of "photosynthesis and respiration, the water cycle, biochemical cycles, food chains and species interactions, soils and aquatic systems, and microbial diversity, and how they respond to key environmental parameters like light, temperature and seasonal changes."[38] In this sense, it represented a style of biological thought and practice closely connected to then ongoing projects at NASA, despite the creators' sympathy with and roots in American counterculture. As we already mentioned, this relationship has been less emphasized in the past than the Biosphere 2's ties to the emergent academic ecology.

The ecologists themselves, however, developed a keen interest in the facility. To Howard Odum, the attraction of Biosphere 2 was that it was large enough "to allow studies of large scale whole-system behavior and of the micro-scale realm where microbes and molecules meet." His brother, Eugene Odum, wrote in *Science* that the Biosphere 2 experiment represented "a new, more holistic level of science." Earlier studies simply did "not capture the essence of whole system responses [at] a scale that will affect humanity." Biosphere 2 apparently moving away from the domination of nature through reductionist technoscience toward a holistic understanding of the world: this resonated with the perspective of some Biosphere 2 insiders. Tony Burgess, the desert biome designer, thought the facility allowed for "a whole new perspective on looking at and managing the ecosystems of the Earth." Biosphere 2's purpose, Poynter argued, was "to bridge the gulf between holistic and reductionist science, a union so desperately needed if we are to grasp and solve the global problems of today."[39]

CONCLUSION

> The colossal cumulative erosion from ten billion annual tourists has seen the fabulously beautiful planet of Bethselamin install a system whereby any net balance between the amount you eat and the amount you excrete while on the planet is surgically removed from your body when you leave; so every time you go to the lavatory there, it is vitally important to get a receipt.
>
> **—Douglas Adams, *The Hitchhiker's Guide to the Galaxy***

IN A LATE 2016 ISSUE OF *WIRED* MAGAZINE OUTGOING PRESIDENT BARACK OBAMA presented his visions of technology and a better future under the symbolic heading "Frontiers." Among other topics the issue addressed space exploration, specifically a new National Geographic television series titled *Mars*. Among the chief concerns of the show, *Wired* reported, was the question "How big does the environmental control and life support system need to be for *x* number of astronauts for *y* number of days?"[1]

That question has been at the heart of efforts to build artificial environments for space travel and habitation for now over sixty years. To Iosef Gitelson and Robert MacElroy, the former directors of their states' respective projects, it was obvious that manmade ecosystems at some point would make "infinitely long human life support outside of the Earth . . . feasible." The size of the system and its carrying capacity, however, they and many others agreed, depended on the development of technologies that were able to return the basics of life back to a crew—that is, it depended on whether and to what extent biological cycles like the food-waste-food loop could be made fully closed and self-sufficient. The knowledge gained after animals and then people were launched into space includes the vital insight that living in space will not be achieved through the building of shining rockets proudly presented by square-jawed astronauts but through sustainable management systems of water, air, food—and not least—sanitation.[2]

Our down-to-earth history of trying to live up in space has described the efforts to build systems that achieved exactly this: closed artificial environments that allowed for human life under inhumane conditions. The Algatron, BIOS-3, CELSS, and Biosphere 2 were major steps on the way toward the scientific understanding and construction of these environments. They were only the beginning, however, of the efforts that are now underway at dozens of institutions occupying hundreds of researchers.[3]

A wealth of experimental approaches exploring what makes environments work emerged in the early space age. The space vehicle came to be understood as an engineered techno-ecology that allowed for short-term excursions, but the actual goal was to build systems that would allow the expected "planetships" to travel into orbit, to the Moon, to Mars, and one day even further out, into the galaxy. Achieving this goal proved more difficult than anybody had thought—going into space was one thing; dwelling in space quite another. It turned out that in addition to technological prowess, a profound change of perspective was required: to appreciate that there is no waste in a closed environment, only nutrients starting their next journey through cycles. Spaceship Earth became a dominant metaphor that has served as an analytical framework to understand the emergence of environmental thinking since the 1970s. Our history argues, in contrast, that within the American and Soviet space programs Spaceship Earth quite accurately described how spaceships had to be designed: namely, like the Earth, as ecosystems with humans as just one component among others and with everything's output being simultaneously every other thing's input.

Space scientists, sanitary engineers, plant physiologists, and nutritionists have spent decades struggling to master life's cycles, and they continue to do so today. Of course, the most elementary version turned out to be incredibly complex. Unicellular algae—small and simple organisms that are highly efficient in photosynthesis—appeared a ready-made solution to waste management, both on Earth and in space. Algae readily digested the carbon dioxide excreted by humans, and conveniently produced oxygen in turn. They even provided some biomass to be consumed. However, the system proved insufficient, and in the course of time, algae-based systems were complemented with and eventually replaced by higher plants, additional microorganisms and macronutrients introduced, time periods prolonged, and levels of self-sufficiency increased. To manage these layers of complexity a variety of physicochemical and bioregenerative technologies were assembled. Scientists and engineers amassed a huge compendium of detail about what systems were needed to support life. In 1974 ecologist Frieda Taub tabulated more than four hundred references to "closed ecological system" at NASA, a figure that in 2018 stands at almost four thousand documents and reports, all freely available on NASA's Technical Reports Server.[4] This American effort was probably dwarfed by the parallel yet still opaque Soviet enterprise. The two Cold War rivals confronted the same problems, and both sought to secure their governments' financial commitment to build spaceships capable of yearlong, decade-long, or generation-long voyages.

This whole project has of course always been controversial, and to some extent our book mirrors the resulting debate. While writing the chapters, we have frequently differed on how to evaluate the general endeavor of living in space. Our history has been informed both by David's techno-ecological optimism and deep fascination with spacefaring adventures and by Kärin's critical view of the expense of going into space, or building the infrastructure to even live there, rather than investing into more sensible and rewarding science on Earth. We both acknowledge, however, that highly significant insights were gathered on the way toward artificial environments. Neither the Soviet nor the American space program has received enough credit for the ecological insight that emerged from their struggle to understand the many feedback loops of closed life-support systems and to contain as well as reuse human waste. Although it is absolutely central, sewage and sanitary engineering is the last thing to be mentioned when utopian space settlements are invoked. As we have described, the designers of the space age envisioned that biomass production, waste processing, and the crew would all be systems approximately equal in importance; the stories told of the space age should reflect that balance as well.

The subject of waste remains if not strictly taboo, at least distastefully weird or childish. It is only occasionally that a refreshingly realistic attitude breaks through. When science-fiction novelist Kim Stanley Robinson was interviewed in 2016 about the latest spacefaring successes of Elon Musk's SpaceX, *Bloomberg News* tentatively raised the issue, what if "someone eccentrically but reasonably asked Musk in Mexico, what's the sewage system like?" "They'll be working to create a closed biological life-support system," Robinson replied, before noting that this was "something that the Soviet Union studied more than the United States during the space race. In these systems, the hope is to waste nothing, so sewage would be integrated back into the ecological system as fertilizer for the farms." Few have done more than Robinson to place waste at the center of space life both in his work for the Mars Society and through his novels. As fans of Robinson will know, algae-based waste systems were integral to his 2015 novel *Aurora*, a story about the journey of a generation ship carrying about 2,100 people on a voyage of 160 years between Sol and Tau Ceti. One of Robinson's characters made the crucial calculation: "You can get two hundred liters of oxygen a week from one liter of suspended algae, if it is lit properly," which was enough for one person to survive.[5] In this novel as well as the actual space programs, the supply of water, air, and food was a massive problem. In Robinson's version, the algae only regenerated oxygen while a separate system processed feces to fertilize plants. In his description of this system, Robinson commented on widespread mechanisms of exclusion and disappearance. The human settlements are in-bounds and visible, while everything below is out of bounds, invisible and inaccessible to the colonists, which is where the waste is going. Only when some characters seek to leave the official colony and live in the out-of-bounds section of the ship does Robinson reveals what is there: the water-recycling system run on chlorella.

Robinson's science fiction replicated science fact to conclude that living in space will be just like living on Earth: it is all about plumbing, gardening, and seeing waste as nutrients. The visions of life among the stars are still there, be they Elon Musk looking to go to Mars via a permanent station in Earth orbit, Mae Jemison working up toward a hundred-year starship, or NASA oscillating between either reviving all the classic hits, including space travels to Mars, or closing down entirely because of lack of funding. Whichever road any of them take, the research of the last half-century has demonstrated that they must equip their ships and stations with algae, bacteria, and living plants. They would do well to recall the insistence of the chief designer of the Soviet space program, Sergei Korolev: that plants were the linchpins to moving humanity into space. Korolev dreamed of "cosmic harvests" in space greenhouses. On top of the nutritional advantage, seeing and feeling green and growing plants seems to offer substantial psychological benefit to the occupants of small orbiting capsules. Cosmonauts Vladimir Lyakhov and Valery Ryumin planted green onions on board the *Salyut* 6 space station in 1979 "more for the soul than science," Valentin Lebedev thought that "without plants, prolonged space explorations are impossible," and Georgy Grechko tried to float into the plant chamber whenever he could for, he later said, mental support. Earth-bound biological cosmonaut Andrei Bozhko was poetic about his experience with plants as he looked back on his yearlong enclosure inside the BIOS-2 facility: "Green plants create a good mood, distract from monotonous and tedious routine matters, and are calming. I am now a staunch supporter of those who believe that the plantation of green plants [will] bring great joy to the crews of space ships and stations. And without fear of exaggeration, I can propose that in space the 'lilac branch' in the space for the human will mean much more for man than on Earth."[6]

There are numerous recent attempts to build closed space habitats. These include, for example, the Hawaiian Space Analogue and Simulation, which has already completed four simulated six-month enclosed missions, while the United Arab Emirates has begun to build a 1.9-million-square-foot domed Mars

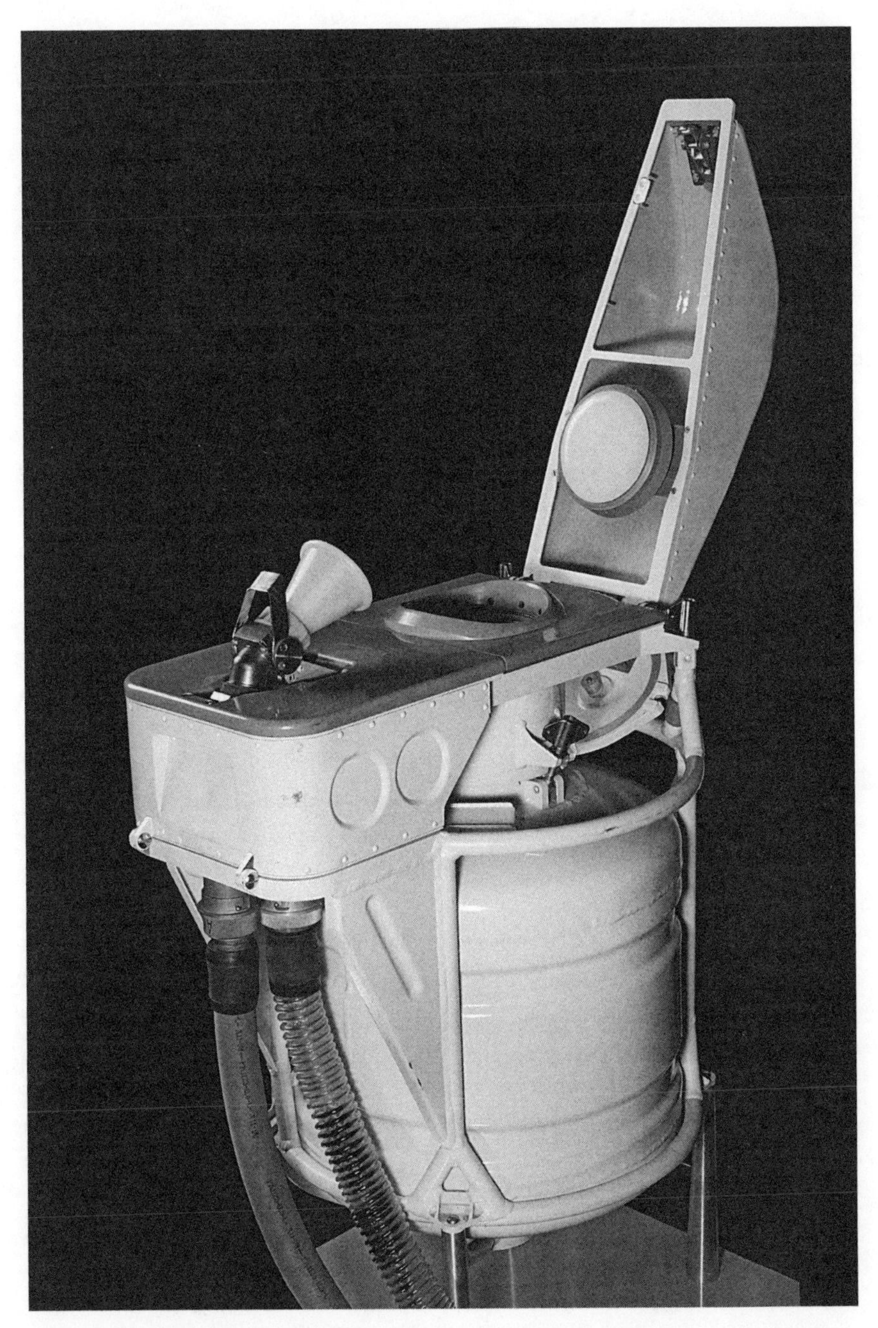

Fig. C.1. The lavatory on *Mir* (female configuration). Now part of the National Air and Space Museum Collection, NASM-A20000786000_PS02.

simulation environment called Masdar City.[7] The technical and social challenge of these facilities will, in all probability, be similar to past simulations. In its efforts to replicate long-term space living, the crew of the Flashline Mars Arctic Research Station experimented with using a new toilet that required special liners to contain the feces dropping from the bowl into an incinerator. After a few weeks they ran out of liners. A crewmember suggested that they all "poop into plastic bags," just like astronauts since the early days of the space program, but Mars visionary Robert Zubrin rejected the idea "because it would stink up the station." Zubrin's crew faced the reality of space living, including the fact that it smells. This, incidentally, was no news to the crews of the ISS, where living in space has slowly become a routine, thanks to internationally cooperative endeavors. One notable success came in 2009, when the European Space Agency worked with the Russian Institute for Biomedical Problems to succeed in repeating a 105-day isolation trial and geared up for a record-setting 520-day study. Still, dealing with waste in space is a preoccupation of all its occupants: among the regular maintenance required by the ISS, on April 24, 2018, astronauts Drew Feustel and Scott Tingle became, NASA said, "space plumbers."[8]

Our book on the pursuit of closed-environment systems for long-duration space missions has pointed to some interesting apparent contradictions. When Apollo 11 went to the moon in 1969, it was the ultimate expression of modernism. Politically and culturally, the glorious aim was launching humans, the pride of God's creation, into space, landing them on the moon. At the same time, the space programs set about preparing for a future life in space on a more permanent basis, and this required them to conceptualize humanity more modestly as just one component of a multispecies, self-sustaining environment. As political support withered, scientists and engineers continued to develop these systems and took care to install a multitude of well-functioning feedback loops. This resulted in some solid ecological studies on the cycles and interconnections between the elements of a closed system, yet, as we have shown, Cold War–era space programs had surprisingly little interest in (or, it seems, use for) contemporary academic ecology. Finally, there was an intense debate over automation versus human agency on board these space stations or spacecraft: as a technical ambition to ensure the project's success and as a funding strategy—there was a time when the idea of a fully automated space station appealed greatly to American politicians, mostly because the Soviet Union did not have one.

The space programs' life scientists and engineers were engaged in both recreating natural environments and constructing artificial systems, and their insights were far-reaching. The ecological implications of miniaturizing an ecosystem had already become clear by the early 1960s, when small space capsules were considered as models of what happens when extreme limits are placed on weight, power, air, water, and food to support life. Research in this direction proved extremely important. Soviet scientists built an entire complex, the BIOS facility, to "balance material cycles," while American sanitary engineers explored "the mutual interdependence of organisms within an isolated environment," from the Algatron to Biosphere 2. This required reconsidering what counted as waste versus nutrients. Looking back on his experience, Mark Van Thillo, the last crewmember to leave the Biosphere 2 in 1993, noted with surprise that "the water we started with in 1991 is still the water inside Biosphere 2. It recycles through the whole system," functioning as "sometimes drinking water, sometimes ocean water, sometimes wastewater." This was exactly what most everyone starring in our book's chapters had been working toward. The closed environments built in the space age sought to integrate humans with plants, animals, and the factors of the non-living environment into highly interconnected networks. Soviet scientists even stated, "Humans should not be regarded as the obligatory (especially sole) heterotrophic link of experimental closed systems."[9] However, Robert MacElroy declared with equal assurance that man was always the center of any environment for space.

Perhaps most importantly, the exploration of life in space opened up discussions about the boundaries of habitability. The participants of closed-environment trials reluctantly had to reassess their assumptions about the place of humans in the environment. In an unexpected twist of events, it was figures within the space program, the epitome of high technology, who first appreciated the scope of the challenge of understanding and controlling the biosphere—long before public awareness was raised in environmental movements. Indeed, notwithstanding the general focus on Mars settlements, a great deal of what these decades of work brought to the fore was about Earth itself. Gitelson and MacElroy concluded in their book, "relations between humanity as a whole and the biosphere rather than human interrelations are becoming critical for the future of both humanity and the biosphere."[10] The question of planetary habitability is distinctly pressing in our Anthropocenic age. It is still an open question whether humanity will be able to learn its lesson and avoid a breakdown. From the history offered here, however, it seems that it might help to embrace what even the military-industrial complex acknowledged as crucial for long-term survival: implementing closed cycles of energy and materials. The engineers and life scientists of the space age built systems where humans subsisted on the waste of plants, while plants fed on the waste of humans. This was the transformative element in Andy Weir's *The Martian*: that excrement is integral to life. It was a goal that scientists at NASA and in the Soviet Union pursued for decades, and which is still unattained. At the same time, perhaps there is a lesson to learn for those who stay on Earth: solutions of our self-made predicament down here will involve staring straight at waste and sanitation. In writing a book about what people have done with their shit in space, we have also written a book that speaks to the problem what people must do with their shit on Earth.

ACKNOWLEDGMENTS

THIS BOOK'S GENEALOGY GOES BACK TO 2015, WHEN ONE OF US (KÄRIN) organized a conference titled "New Perspectives for the History of the Life Sciences" and the other (David) submitted an abstract on something that sounded genuinely weird, in fact, like a hoax: namely, phytotrons, greenhouses with completely controlled environmental parameters. A brief search on the internet confirmed, however, that this was a real thing, and David was invited. It then happened that Kärin, who was working at the time on the history of photosynthesis and algae cultures, stumbled across a particularly bizarre exemplar of this family of controlled environments, which was the Algatron. We started to dig further into the sources, and the next thing we knew we were writing a book together on the history of bioregenerative life-support systems in space, a topic we found everywhere in the sources and almost nowhere in the literature. Although—or even because—our approaches to style and argument were very different, this became a most joyful and satisfying experience of complementary expertise and mutual learning. We hope you readers are as happy with the result as we are.

Along the way, many people have contributed their thoughts, commentaries, advice, expertise, and sources to the project, or provided practical support, all of which we are deeply grateful for. John Benemann, from MicroBio Engineering Inc. and the Algae Biomass Organization, shared some terrific stories and insights from the world of industrial algae with us, as did Tryg Lundquist from the same company, who once had closely worked with William Oswald, one of the inventers of the Algatron. Ray Wheeler of NASA's Kennedy Space Center and Dan Barta of NASA's Johnson Space Center also supported our project in most generous ways. We are furthermore deeply grateful to Lloyd Ackert for his invaluable translations of Bozhko's book, which helped enrich the Soviet chapter so much. Archives, libraries, and reading rooms are of course fundamental for all historical work, ours included. Particular thanks, however, go to the Slavonic Studies unit of the British Library for help in tracking down sources, the John Jay Library staff, the staff at the National Archives, the staff of the University Library of LMU Munich and the Bavarian State Library. We also had the privilege to receive prompt, diligent, and much-appreciated assistance of various helpful souls, including Magnus Altschäfl, Maximilian Heumann, Marina Schütz, Claus Spenninger, Laurenz Denker, and, in particular, Merlin Wassermann.

Finally, thanks go to the various reviewers and editors that helped us improve our work and, eventually, have it published in much better shape. This includes the journal *History and Technology*, most especially Amy Slaton and Tiago Saravia, who accepted a paper of ours in which we developed some of our thoughts in embryonic form, and, of course, the University of Pittsburgh Press for their sage advice and continuous helpful guidance. To Abby Collier and Amy Sherman, our wonderful editors at Pittsburgh, Alex Wolfe, editorial and production director, and Sarah C. Smith, our keen-eyed copyeditor, our many thanks for all your work, advice, and efforts to squeeze in another dozen illustrations.

FROM DAVID P. D. MUNNS

When I was a kid, I wanted only one type of toy for birthdays and Christmases: classic space LEGO. By the time I was about twelve (mid-1980s), I owned nearly every space LEGO set available, and bits of many others. When, years later, *The Lego Movie* came out, I was overjoyed to see the return of Benny the spaceman, and, now equipped with slightly more pocket money, ran out to the LEGO Store at Rockefeller center and bought the re-release of a classic space LEGO spaceship. I spent a glorious hour building it that very night. Safe to say, my childhood LEGO obsession probably played some part in the decision to write a book about controlled environments for space. All LEGO spacemen (they were all men) came equipped with a helmet and oxygen tanks, though I do not recall any idea that the oxygen must have come from somewhere or that there was carbon dioxide that needed to be processed in turn. The idea that a spaceman would more realistically wear a suit more like the "stillsuits" of Frank Herbert's

Dune, in which one would recycle waste back into drinkable water, was not yet an idea I had glimpsed—it would wait until John Apperson insisted I read *Dune* in 2001. Ah, what a simple time it was before all such ecological concerns started weighing in on my childish imaginary worlds. Realistic maturity now insists that air, water, food, waste, and fuel cannot just be a one-way extractive process but must be a part of a cyclical material economy. We must return all we take out because we all live inside in the same closed biological economy.

This work, like the efforts it described, has been interdisciplinary, interinstitutional, and collaborative. I can safely say that Kärin Nickelsen is the best coauthor one could hope for, as her Teutonic comments are as lashingly helpful as her knowledge about the history of science is flamingly brilliant. To facilitate our work, I spent a glorious and rewarding semester at the Rachel Carson Center in Munich, which for over a decade has been a critical center and a leading light investigating the environmental humanities. There among the sterling resources and exceptional intellects of my fellows, the directors of the RCC, Helmuth Trischler and Christoph Mauch, have created the kind of scholarly enthusiasm that one might believe exists only somewhere over a rainbow; I can assure everyone that it is in Munich. Positive and constructive advice flowed from my fellows at the Rachel Carson Center's "Works-in-Progress" sessions, a model for academic community, the weekly colloquiums, and regular visitors through the center.

In addition, I owe many thanks to my institutional home, the John Jay College of the City University of New York, and our (now-departed) provost Jane Bowers for supporting scholars in their research both with release time and much appreciated funds. To my chair and friend, Allison Kavey, I say simply thank you and hope you appreciate how important you have always been and will always remain: we are Brian and Michael at Babylon forever.

Across institution, nations, and disciplines, academic work always seeks new audiences, new perspectives, and new challenges. I would like to acknowledge the comments, questions, and advice gathered from the various audiences who have listened to a book about sanitation in space and offered, always, insightful improvements. Thanks notably to Michael Robinson and his terrific blog, *Time to Eat the Dogs*, as well as the Rachel Carson Center's blog staff for their work on *Seeing into the Woods*. My thanks to Valerie Neal for leading our great session at a SHOT meeting. Likewise, the people gathered at the meetings of the History of Science Society, Columbia History of Science group, the History Colloquium at the National Air and Space Museum, the Crafting the Long Tomorrow symposium at the Biosphere 2, and the Society of the History of Technology as well as the immersive experiences of the Rachel Carson Center's Lunchtime

Colloquium series, the "Again, Method" workshop on the history and philosophy of biology at LMU, and the Deutsches Museum Montagskolloquium have all been foundational in shaping this work.

Being stuck for the moment on this tiny blue rock as we all are, we can take comfort in sharing our short existence with the partners in one's work who don't get the glory of a footnote but without whom little would taste as sweet: Diane Kagoyire, J. J. Shirley, Catherine Jackson and Andy Warwick, Luis Campos, Colin Milburn, Brad Bolman, the "Yes" Appersons, the Wisnioskis, the Windeyers, the Woods, the Borises, the Griffins, the DeLeons, A. J. Benitez and Brad Oister, Kenneth Moore and Derek Bishop, Eric Kolb, Scott Knowles, Dara Byrne, Michael Pfeiffer, Matt Perry, Hyunhee Park, Sara McDougall, Adam Berlin, and Walter Fralix, all helped in more ways than they know. Joseph DeLeon quite simply completes my world. Thanks finally to my parents, Peter G. and Susan Munns, and to Lillian, Max, Trudi, and David MacKay for their patience with their son/brother/uncle's continuing wanderings.

Thanks finally to Carl Sagan's *Cosmos*, which helped begin my journey.

FROM KÄRIN NICKELSEN

There is only so much space and so many words that you can spend on the acknowledgment section of a book, and since David, for once, refused to have his prose cut short, I will have to be brief. It helps that I have no childhood memories of LEGO spacemen to offer and have never been particularly excited about astronauts or science fiction. It was more a growing fascination for the fact that people spent so much time, thought, and energy on designing life support systems for space, which dragged me into the subject of this book. In addition, I found it odd, and in urgent need of correction, that algae and life scientists were (and still are) so important for NASA's adventures into space but have gained so little prominence among historians of science and technology. Finally, it has been immensely enjoyable and rewarding to work with David Munns, both on a personal and intellectual level. His writing speed and enthusiasm are stunning, as is his ability to unearth the most intriguing sources. The letters by a German dentist, who innocently offered his expertise in decomposing organic materials to the service of the heads of NASA, is only one of the many archival gems that David discovered in the archives.

I would like to thank the Department of History at LMU Munich for granting me a sabbatical to work on the book, and I extend my thanks to dear colleagues, friends, and family, but will not elaborate (see above). Neither will I repeat the various venues and audiences where the project was presented earlier and who offered helpful feedback and criticism. Let me add to this list, however, my own history of science seminar at LMU Munich, mostly graduate students and some postdocs. When David was at the RCC, we discussed two chapters of our book with them. Unimpressed by academic decorum, the students tore the draft into pieces, deservedly so, and demanded that we sharpen our claims and strengthen the argument. I am deeply grateful for the pleasure and privilege of working with this group of bright, aspiring, and immensely likable young scholars, and I hope that the final version will find their approval. Lastly, my two sons have accompanied the book from the very beginning and have learned more about shit in space than they ever had wanted to. This book is for them.

LIST OF ABBREVIATIONS

THE COLD WAR ERA IS KNOWN FOR ITS MYRIAD ACRONYMS. WHEREVER POSSIBLE we have kept their usage to a minimum but a list is inevitably necessary.

BSRP	Biological Systems Research Program (NASA)
CDG	Concept Design Group (NASA)
CELSS	Controlled Ecological Life Support System
GARDEN	Growth Apparatus for the Regenerative Development of Edible Nourishment
ISS	International Space Station
LMLSTP	Lunar-Mars Life Support Test Project
MOL	Manned Orbiting Laboratory (NASA, USAF)
NAS	U.S. National Academy of Sciences
NASA	National Aeronautics and Space Administration
STS	space transportation system (space shuttle)
USAF	United States Air Force

Note on Units. While the United States used the imperial units of feet and pounds for their weight and measures, the Soviet Union adopted the metric system of meters and kilograms. We use both in their associated contexts, though it must be noted, as one of our old physics professors used to say, one really shouldn't use wheelbarrow per square banana units, only metric.

NOTES

Introduction

1. See the official announcement of October 12, 2016, on NASA's website at http://www.nasa.gov/feature/space-poop-challenge. "The Space Poop Challenge," NASA Johnson Space Center, https://www.nasa.gov/sites/default/files/atoms/files/space_poop_challenge_summary_and_winners.pdf.

2. Weir, *The Martian*, 14.

3. One of the first stories about a generation ship, from 1940, uses a six-hundred-year journey. Wilcox, "The Voyage," 160. Population control as an environmental safeguard is also mentioned (161). See Robinson, *Aurora*.

4. Walker and Granjou, "MELiSSA the Minimal Biosphere," 59–69. The report of July 24, 2017, "ESA's MELiSSA Life-Support Programme Wins Academic Recognition," celebrating Lasseur's Doctor Honoris Causa degree from the University of Antwerp, is available at ESA's website at http://www.esa.int/. Wheeler, "Bioregenerative Life Support," 41–67; Hinghofer-Szalkay and Moore, "Some Comments," 542–61; Gitelson et al., *Manmade Closed Ecological Systems*.

5. On the early Cold War period, see, e.g., Gaddis, *The United States and the Origins*; Gordin, *Red Cloud at Dawn*. For the domestic character of the Cold War, see, e.g., Kirkby and Scalmer, "Social Movement, Internationalism and the Cold War," 3; Craig and Radchenko, *The Atomic Bomb*, x.

6. See, e.g., Paglen, *Blank Spots on the Map*. Isaac and Bell, *Uncertain Empire*; Leffler and Westad, *Cambridge History of the Cold War*; LaFeber, *America, Russia, and the Cold War*; Ambrose and Brinkley, *Rise to Globalism*; Brands, *American Dreams*; Adas, *Dominance by Design*, 18; McDougall, . . . *The Heavens and the Earth*.

7. Lilienthal, *Change, Hope, and the Bomb*, 73; Siddiqi, *Challenge to Apollo*, xi; Richers, "Remembering the Soviet Space Program," 843–47.

8. Mailer, *Of a Fire on the Moon*, 19. See also Wolfe, *The Right Stuff*; Michener, *Space*; Degroot, *Dark Side of the Moon*; Burrow, *This New Ocean*. More than half the chapters are devoted to propulsion in Mallove and Matloff, *The Starflight Handbook*. See also Portree, *Humans to Mars*. Such works parallel classics of the automotive history genre such as Ingrassia, *Engines of Change*.

9. Shetterly, *Hidden Figures*; Holt, *Rise of the Rocket Girls*. Rigidly defined gender roles meant that "at a very basic level, it never occurred to American decision makers to seriously consider a woman astronaut." Weitekamp, *Right Stuff, Wrong Sex*, 3. Rigid concepts of race mean similarly that nonwhite astronauts were never seriously considered in the early space program. See Phelps, *They Had a Dream*.

10. Roach, *Packing for Mars*; Olson, *Into the Extreme*, 168; Benford and Zebrowski, *Skylife*, 15; Ard, "Garbage in the Garden State," 57–66; Douglas, *Purity and Danger*. For the social history turn that privileged the lowly or the forgotten in our histories, we are indebted to Thompson, *Making of the English Working Class*, rescuing the poor English stockinger.

11. Duca, "Nutrition-Waste Complex," 9.

12. Kelly, *Endurance*, 52. Astronaut Tim Peake said exactly the same thing. Peake, *Ask An Astronaut*, 88.

13. See Table 1, from Slavin and Oleson, "Technology Tradeoffs," 402. On physicochemical systems, see Eckart, *Spaceflight Life Support*; and Taub, "Closed Ecological Systems," 140. There are hints that physicochemical methods were tainted by their military use in submarines and aircraft. Gitelson et al., *Manmade Closed Ecological Systems*, 34; Eckart, *Spaceflight Life Support*, 80–81.

14. Bowman, "Food and Atmosphere Control Problem," 159–97; comments by Sharon Skolnick at the Space Station Human Factors Research Review: Architecture Panel Discussion, December 3–6, 1985, in *Space Station Human Factors Research Review*, vol. 3 (NASA: Scientific and Technical Information Division, 1987), 191–210, 197, Box 3, Series: R&D Projects—Space Station Advanced Development Program, 1985, RG255, NARA.

15. Oswald and Golueke, "Environmental Control Studies," 184; MacElroy and Averner, "Space Ecosynthesis," 8–9.

16. See the essays in Cronon, *Uncommon Ground*. Allen, *The Quest*, 218.

17. McCray, *The Visioneers*; Olson, *Into the Extreme*, 168. Anker, "Ecological Colonization of Space," 239–68, described these visions for architects, while Messeri, *Placing Outer Space*, 17–18, turned to recent model Mars colonist missions and those searching for exoplanets. See also Zubrin, "Significance of the Martian Frontier"; and Duggins, *Trailblazing Mars*, ch. 7.

18. Chekhonadskiy, "Cybernetics and Space Biology," 192; Mody, "Square Scientists and the Excluded Middle," 63–64; Egan, "Survival Science," 26–39; Kaiser and McCray, *Groovy Science*.

19. Sharpe, *Living in Space*, 77; Shepelev, "Some Aspects of Human Ecology," 167; Olson, *Into the Extreme*, 11, 75.

20. Astrobiology is the study of the origin and evolution of life on Earth and the possible variety of life elsewhere in the cosmos. Catling, *Astrobiology*, 2. A prominent answer to the question of what is life was given by Schrodinger, *What Is Life?*. The uncertainty of any

definition is described by Folsome, *Origin of Life*, ch. 6. See also the essays in Scientific American, *The Biosphere*. For the perspective of astrobiology, see Dick and Strick, *Living Universe*, 4, 67–71.

21. Contemporary surveys of the range of issues for "space medicine" include Allen, *The Quest*, 199; Sharpe, "Human Parameter." Bjurstedt, *Proceedings of the First International Symposium*, 473–92 (Reynolds presented at this conference, as did Eugene Konecci); Kelly, *Endurance*, 18; Human Research Program, NASA, https://www.nasa.gov/.

22. Quoted in Committee on Science and Astronautics, *Astronautical and Aeronautical Events of 1962*, 78; Munns, *Engineering the Environment*, 214; Gitelson et al., *Manmade Closed Ecological Systems*, 9.

23. Poole, *Earthrise*, 152; Höhler, *Spaceship Earth*, 16–18; Taub, "Closed Ecological Systems," 139, 142; Aronowsky, "NASA and the Dream," 366–371. On the Odums' contract being terminated, see Taub, "Closed Ecological Systems," 153.

24. Kirk, *Counterculture Green*, 8; Rome, *Genius of Earth Day*; Egan, *Barry Commoner*, 84; Egan, "Survival Science," 26–39. Fred Turner describes how the counterculture began as an anti-technological movement in Turner, *From Counterculture to Cyberculture*.

25. Trim, "A Quest for Permanence." Other examples include Stewart Brand and J. Baldwin's "Alloy" project and Drop City; see Kirk, *Counterculture Green*, ch. 3. In the famous volume *Space Colonies*, edited by Stewart Brand, Lynn Margulis was offered the chance to comment on Gerard O'Neill's conception of space colonies and commented that the "John Todd's of the World (e.g. holistic biological thinkers and doers) must connect with O'Neill and his crew to help stop the handwaving." Brand, *Space Colonies*, 35.

26. For all-too-brief introductions, see Birnbaum and Fox, *Sustainable Revolution*, 268; Dawson, *Ecovillages*, 44.

27. Kirensky et al., "Theoretical and Experimental Decisions," 79; Clarke, *Promise of Space*, 292.

28. Jesco von Puttkamer (NASA's Office of Space Flight), "Human Role in Space," NASA Space Station Concept Development Group, Habitability Workshop, August 30–31, 1983, in "NASA Space Station Task Force Concept Development Group," vol. 6, Box 4, Series "R&D Projects—Space Station Program Reports, 1982–1984" (acc. no. NRHS255-09-001), RG255, NARA.

29. Worster, *The Good Muck*, 4–5. Howell, "Fecal Matters," 137–51. An example of bowdlerized environmental history is David Soll, *Empire of Water: An Environmental and Political History of the New York City Water Supply* (Ithaca: Cornell University Press, 2013), which is marvelously detailed on the evolving story of how New York City consumes nearly a billion gallons of water per day but says almost nothing about the outflow of that same billion gallons.

30. Köster, "Waste to Assets," 170–171. For waste studies, see Hamblin, *Poison in the Well*; and Siniawer, *Waste*. In the United States, new patterns of consumption reshaped the lives of American Cold War families. See May, *Homeward Bound*, esp. ch. 7; and Cohen, *A Consumer's Republic*. Moreover, consumption and its attendant waste are now understood as major causes and consequences of climate change, see Klein, *This Changes Everything*; Guha, *How Much Should a Person Consume*.

31. McNeill, *Something New under the Sun*, 288. See Sellers, *Hazards of the Job*; Murphy, *Sick Building Syndrome*.

32. Galbraith, *The Affluent Society*, 253; "The Gold Violin," *Mad Men*, season 2, episode 7, written by Jane Anderson, Andre Jacquemetton, Maria Jacquemetton, and Matthew Weiner. In the same episode, the advertising staff try to put together a campaign to sell Pampers diapers, which, while expensive, have one key advantage: "you get to throw them away."

33. Worster, *Good Muck*. This is significant, given the claim that we might learn a lot from what people consider waste. See, e.g., Laporte, *History of Shit*; Simons, "Waste Not, Want Not," 73–98.

34. Tarr, "From City to Farm"; Tarr et al., "Water and Wastes," 226–63; Melosi, *Sanitary City*; Hoagland, *The Bathroom*; Reid, *Paris Sewers and Sewermen*.

35. Wynn, "Foreword," vii; Benidickson, *Culture of Flushing*; Peake, *Ask An Astronaut*, 3; Pogue, *How Do You Go to the Bathroom in Space?*, 71.

36. Simons, "Waste Not, Want Not," 73–98; "Request for Major Capital Improvement Sanitary Engineering Research Laboratory," October 1957, CU149, box 40, folder 16, Archives, Bancroft Library, University of California, Berkeley; Rome, *Genius of Earth Day*, 19; Melosi, *Sanitary City*, 180–201; Van der Ryn, *Toilet Papers*, 34.

37. Rome, *Genius of Earth Day*, 47–56; Rome, "Give Earth a Chance"; Egan, *Barry Commoner*, 2–3; Gottlieb, *Forcing the Spring*, 34–36; Kirk, *Countercultural Green*, 21–27.

38. Gitelson et al., *Manmade Closed Ecological Systems*, 25; Kelly, *Out of Control*, 132–33; Mendell, "Space Exploration Initiative," 328; Sagan, *Biospheres*, 20.

39. A remarkable source for this book, and any other topic related to NASA, is the NASA Technical Reports Server, ntrs.nasa.gov, which boasts more than 1.2 million documents from the space program. The database is fully searchable and also contains a host of translations of Soviet documents from the open literature. A wealth of archival material is available through the National Archives listed under record group 255. The papers of the life science directorate of NASA archives are stored in San Bruno, California, but important materials turn up in many places, most remarkably for this story in the H. Guyford Stever papers at the Gerald R. Ford presidential library in Ann Arbor Michigan.

40. Gitelson et al., *Manmade Closed Ecological Systems*, 3.

Chapter 1: When America Aimed beyond the Moon

Epigraph: "Space Quotes," vol. 2, no. 3, March 1964, Folder "Correspondence with NASA Headquarters, Oct 18, 1963," Box 2, Series 36: Life Sciences Directorate, 1963–67, RG255, NARA (acc. no. 255-93-022).

1. Max Gunter, "The American Scientist: Man or Superman?" *Saturday Evening Post*, December 16, 1969, 30; Ford, *Cult of the Atom*, 236. For broader discussions see Adas, *Dominance by Design*, 18, 225; Jervis, "Identity in the Cold War," 25.

2. Siddiqi, *Sputnik and the Soviet Space Challenge*; Dickson, *Sputnik*; quoted in Daniel J. Kevles, *The Physicists*, 385.

3. Low, "Biological Payloads," 313–14. There is some doubt now if Laika lived for six or seven days, but the Americans at the time believed she had, so the flight of Laika continued to be a touchstone for years afterward. NASA's director of biotechnology and human research, Eugene Konecci, noted in 1963 that Laika still held the record for the longest time in space for a living thing. Eugene B. Konecci, "Bioastronautics Review—1963," in Folder "Biotechnology & Human Res. Advis. Cte., 7/19/63," Box 1, Series 36, "Life Sciences Directorate, 1963–67,"

RG255, NARA (acc. no. 255-93-022), 12. For Laika and her legacy, see Turkina, *Soviet Space Dogs*, 89; and Wellerstein, "Remembering Laika."

4. Wieland, *Designing for Human Presence in Space*, 102; Burgess and Dubbs, *Animals in Space*, 127–29; Low, "Biological Payloads," 313–14.

5. Stevenson quoted in Matusow, *Unraveling of America*, 12. On the Kitchen Debate see May, *Homeward Bound*; Cohen, *A Consumer's Republic*, 126. See also Oldenziel and Zachman, *Cold War Kitchen*.

6. On the transformations to high school curriculums, see Rudolph, *Scientists in the Classroom*. For colleges and universities, see Kaiser, "Cold War Requisitions"; Leslie, "Playing the Education Game to Win." In general, Kaiser and Warwick, *Pedagogy and the Practice of Science*.

7. McDougall, . . . *The Heavens and the Earth*; Wolfe, *Competing with the Soviets*, 95; Weitekamp, *Right Stuff, Wrong Sex*, 3–4, 160–62. For a brief biography of Tereshkova and an analysis of her impact, see Sylvester, "She Orbits over the Sex Barrier," 195–212.

8. Konecci, "Bioastronautics Review—1963," 13. Voskhod I was launched on October 12, 1964. Quoted in Brooks and Ertel, *Apollo Spacecraft*, 348; Science Policy Research Division, *Review of the Soviet Space Program*, 47.

9. "Space Quotes." It was said by no less than Senator Stuart Symington. Klerkx, in *Lost in Space*, 13, is one of the few who note that one of the justifications of Apollo was as proof of concept for future permanent human settlements.

10. Document III-2, in Logsdon et al., *Exploring the Unknown*, 1:404; National Aeronautics and Space Administration, *Challenge of Space Exploration*, 42. See also Portree, *Humans to Mars*.

11. Muenger, *Searching the Horizon*.

12. Konecci, "Bioastronautics Review—1963."

13. Kirensky et al., "Theoretical and Experimental Decisions," 75.

14. Jesse Orlansky, "Bioastronautics R&D in NASA," July 12, 1963, in folder "MIT Scientific Advisory Board–Bioastronautics," Box 35, H. Guyford Stever Papers, Gerald R. Ford Library.

15. Minutes of Meeting of NASA Research Advisory Committee on Biotechnology and Human Research, September 9, 1963, Folder "Biotech & Human Res. Advis. 7/19/63," Box 1, Series 36: Life Sciences Directorate, 1963–67, RG255, NARA (acc. no. 255-93-022).

16. Memo from H. Julian Allen, July 1, 1966, Folder "Life Science Accomp. Dec 6, 1966," Box 2; and final comments by Ralph Stone on the Langley Research Center Life Support Studies, in Folder "Biotechnology & Human Res. Advis. Cte.,' Box 1, Series 36: Life Sciences Directorate, 1963–67, RG255, NARA (acc. no. 255-93-022). Dick and Strick, *Living Universe*, 37. Exobiology grew rapidly through the 1960s, gathering some seventy researchers across three branches, chemical evolution, biological adaptation, and life detection. Biochemist R. D. MacElroy, who played a prominent role in the development of NASA's Controlled Environment Life Support System, CELSS (chapter 4), was already part of the Biological Adaptation Branch in 1970. See Dick and Strick, *Living Universe*, 39.

17. Acevedo, "In Memoriam Dr. Harold P. Klein," 549–51; quoted in David, "Political Acceptability of Mars Exploration," 36.

18. Konecci, "Bioastronautics Review—1963"; Clarke, *Sands of Mars*; Clarke, *Promise of Space*, 124.

19. Mailer, *Of a Fire on the Moon*, 132.

20. Within the history of ecology, there is a debate over the extent to which ecology was either handmaiden or a respected science that could offer important knowledge to the Cold War atomic state about radiation. For the handmaiden and salve, see Smith, *A Peril and a Hope*, 77. In contrast, Bocking, *Ecologists and Environmental Politics*, 68. More correctly they were not one or the other but both, as Angela Creager argues about radioactive tracers. Tracers are better understood as a "dual" technology. Creager, *Life Atomic*, 19. Also Hagen, *An Entangled Bank*; Curry, *Evolution Made to Order*.

21. See the collection of essays in Westwick, *Blue-Sky Metropolis*. The aerospace industry remains mostly hidden behind what Trevor Paglen calls the "black world" of secret organizations. Paglen, *Blank Spots on the Map*, 172. See also his Google presentation by the same name, available on YouTube.

22. Lorne Proctor, "Life Sciences—Advanced Concepts," attached to a letter to Harold Klein, March 23, 1966, Folder "Proposals, June 20 1966," Box 2, Series 36: Life Sciences Directorate, 1963–67, RG255, NARA (acc. no. 255-93-022).

23. Compton and Benson, *Living and Working in Space*, 152. Echoes and repetitions of this standard argument occur in Kevles, *Almost Heaven*, 51; Pitts, *Human Factor*, 114–15. Important background to NASA's attempts to restart much of the abandoned efforts into "waste management subsystems" and "man-machine integration" can be found in "The Life Sciences Program of the National Aeronautics and Space Administration," February 1, 1974, Document III-19, in Logsdon et al., *Exploring the Unknown*, 6:376.

24. Mating rats at g>3.5 was evidently challenging. Hartman, *Adventures in Research*, 481. Other examples of this historiography abound. In his summary of NASA's ideas about space stations, Roger Launius mentions that plans for space station were pushed to the bottom of NASA's priority list. While strictly true, this overlooks the substantial efforts in the life sciences that underpinned a space station. Launius, "Space Stations for the United States," 541. This has become the accepted story; see Alice Gorman, "The Sky is Falling," 529.

25. Klerkx, *Lost in Space*, 333.

26. Colin Pittendrigh, "Preface," in Pittendrigh et al., *Biology and the Exploration of Mars*, viii; Konecci, "Closed Ecological Systems," 3–20; Belasco and Perry, "Waste Management and Personal Hygiene."

27. "Minutes of Meeting NASA Research Advisory Committee on Biotechnology and Human Research," December 6–7, 1965, Folder "Biotech & Human Res. Cte. Mtg 12/6–7/65," Box 1, Series 36: Life Sciences Directorate, 1963–67, RG255, NARA (acc. no. 255-93-022), 8. Also Konecci, "Closed Ecological Systems," 5–9; Wilkins, "Man, His Environment and Microbiological Problems," 8.

28. Weir, *The Martian*, 9; Salisbury, "Controlled Environment Life Support Systems (CELSS)," 171; Seedhouse, *Martian Outpost*, 144. According to Robert Zubrin, while the Saturn V could launch weight at about $7,000 per kilogram, the actual cost to launch weight in the space shuttle was closer to $20,000 per kilogram. Zubrin, *Mars on Earth*, 5–6.

29. Orlansky, "Bioastronautics R&D in NASA."

30. The economics of cost (weight) versus time for various life support configurations. From Cooke, "Ecology of Space Travel," 499.

31. Prime examples include Cooke, "Ecology of Space Travel"; Anker, "Ecological Colo-

nization of Space," 239–68; Messeri, *Placing Outer Space*; Poole, *Earthrise*. It is odd because at least one ecologist noted the absence of the ecologists: Taub, "Closed Ecological Systems," 141, 151–52. Peder Anker shows, however, how the ecologists shaped ideas of space colonization after the 1970s and in turn shaped environmental thinking about sustainable and maximum human populations on Earth. Valerie Olson's book fixates on the design of closed-style habitats rather than the engineered interior, but also too readily accepts current participants' stories. For example, Olson notes that BIO-Plex semi-closed ecological system at the Johnson Space Center "has yet to be occupied" "like other aspirational" systems, which is not true of either NASA or its contractors. We count at least seven built and tested closed-environment systems since 1962. Olson, *Into the Extreme*, 144.

32. Stanley Deutsch, "Human Factor Systems," January 5, 1964, Folder "Correspondence with NASA Headquarters, Oct 18, 1963," Box 2, Series 36: Life Sciences Directorate, 1963–67, RG255, NARA (acc. no. 255-93-022), Figure 1, "Biotechnology & Human Research Program, September 1962."

33. "Minutes of Meeting NASA Research Advisory Committee on Biotechnology and Human Research," 9. On the cybernetic zeitgeist, see Pickering, *Cybernetic Brain*; and Bocking. *Ecologists and Environmental Politics*, 74.

34. Shepelev, "Some Aspects of Human Ecology," 167; Hark, *Star Trek*.

35. Konecci, "Bioastronautics Review—1963." Emphasis in original. What became the sensational section of his review was the few minutes Konecci spent talking about some Russian experiments with "thought transference over distance," even building some kind of "electronic apparatus" to "probe and control such brain-mind mysteries as energy transfer phenomena, or 'biological radio communication.'" (13). Konecci's remarks on "mind reading," as NASA understood them, resulted in an official censure. Subsequently, senior NASA administrators moved Konecci sideways in the organization as they questioned his "scientific maturity." Memorandum from the Consultant for Life Sciences to the Administrator to the Associate Administrator, October 4, 1963, Document III-10, in Logsdon et al., *Exploring the Unknown*, 6:347.

36. Minutes: NASA Research Advisory Committee on Biotechnology and Human Research, March 11–12, 1963, in Folder "Biotechnology & Human Res. Advis. Cte., 7/19/63," Box 1, Series 36: Life Sciences Directorate, 1963–67, RG255, NARA (acc. no. 255-93-022). Stanley Deutsch, "Human Factor Systems." Emphasis in original.

37. Steinbuch, "Man or Automaton," 476; Nelson, *Pushing Our Limits*, 185.

38. Minutes of Meeting of NASA Research Advisory Committee on Biotechnology and Human Research, September 9, 1963.

39. Press Release, November 16, 1964, "Keynote Address by Dr. Edward Welsh," Folder "Biotech & Human Res. Comm. 9–11–64," Box 1, Series 36: Life Sciences Directorate, 1963–67, RG255, NARA (acc. no. 255-93-022).

40. John Billingham to Walton Jones, September 8, 1966, Folder "NASA Manned Space Station, Oct 11, 1966," Box 2, Series 36: Life Sciences Directorate, 1963–67, RG255, NARA (acc. no. 255-93-022).

41. Walton L. Jones to Harold Klein, September 23, 1966, Folder "Headquarters Correspondence," Box 2, Series 36: Life Sciences Directorate, 1963–67, RG255, NARA (acc. no. 255-93-022).

42. See the centerfold in General Electric Company, Missile and Space Division, *Challenge* 3, no. 2 (1964): 24–25, in Folder "MIT-Scientific Advisory Board-Manned Orbiting Laboratory," Box 37, H. Guyford Stever Papers, Gerald R. Ford Library.

43. Bushnell, "Beginning of Research in Space Biology," 39–49.

44. Launius, "'Astronaut Envy?,'" 61–78.

45. Quote "determin[e] the military usefulness of man in space," from "Platform for Progress," General Electric Company, Missile and Space Division, *Challenge* 3, no. 2 (1964): 12, in Folder "MIT-Scientific Advisory Board-Manned Orbiting Laboratory," Box 37, H. Guyford Stever Papers, Gerald R. Ford Library. The article further notes that General Funk's wife was an "attractive" "young starlet" from Hollywood (10). Other quotes from "Excerpts from DoD New Release No. 1556-63, 10 Dec 1963," in Folder "MIT-Scientific Advisory Board-Manned Orbiting Laboratory," Box 37, H. Guyford Stever Papers, Gerald R. Ford Library.

46. Clarke, *Promise of Space*, 125; *New York Times*, July 1, 1967, p. 1. See also Maj. Robert Lawrence, "Secret Astronaut," PBS, https://www.pbs.org/wgbh/nova/astrospies/prof-08.html; "Maj. Robert H. Lawrence: The First African-American Astronaut Designee," in Phelps, *They Had a Dream*, 47–76.

47. See Spigel, "White Flight," 47–71; Peake, *Keeping the Dream Alive*, 264–65.

48. Mailer, *Of a Fire on the Moon*, 125. See also Höhler, *Spaceship Earth*, 59; Tibbe, *No Requiem for the Space Age*, 36–37; *New York Times*, June 11, 1969, p. 1. Lawrence himself had been tragically killed in December 1967 during a training exercise with the Lockheed F-104 Starfighter.

49. NAS Press Release, November 17, 1964, Folder "Exobiol. Exec. Summer Study Aug 12, 1964," Box 2, Series 36: Life Sciences Directorate, 1963–67, RG255, NARA (acc. no. 255-93-022).

50. Hess to Webb, October 30, 1964, attached to "Statement of the Space Sciences Board of the National Academy of Sciences on National Goals in Space, 1971–1985," October 28, 1964, Folder "Exobiol. Exec. Summer Study Aug 12, 1964," Box 2, Series 36: Life Sciences Directorate, 1963–67, RG255, NARA (acc. no. 255-93-022). NAS Press Release, November 17, 1964.

51. Hess to Webb, October 30, 1964; Space Station Requirements Steering Committee, *The Needs and Requirement for a Space Station,* September 1966, Folder "NASA Manned Space Station, Oct 11, 1966," Box 2; and Memo from RBM/Exec. Sec., Rockefeller Archive Center, on BHR to director, Biotechnology and Human Research, January 5, 1966, Folder "Biotechnology & Human Res. Advis. Cte., 12/6–7/65," Box 1, Series 36: Life Sciences Directorate, 1963–67, RG255, NARA (acc. no. 255-93-022).

52. Minutes: NASA Research Advisory Committee on Biotechnology and Human Research, March 11–12, 1963; W. Hypes, "Integrated Regenerative Life Support Systems," report attached to Memo from RBM/Exec. Sec., Rockefeller Archive Center, on BHR to director, Biotechnology and Human Research, January 5, 1966, 17.

53. J. J. Konikoff, "Closed Ecologies for Manned Interplanetary Flight," Report R63SD83, Aerophysics Section, Space Sciences Laboratory, Missile and Space Division, General Electric, October 1, 1963, 2. Lockheed also manufactured the Agena rocket, used for NASA but also the military. Brzezinski, *Red Moon Rising*, ch. 6. Any NASA funding was still an order of magnitude below the defense contracts given to aerospace industries. See, e.g., Hughes,

Human-Built World, 81; and for the development of the intercontinental ballistic missile, Hughes, *Rescuing Prometheus*, ch. 3.

54. Jagow and Thomas, "Study of Life Support Systems," 77.

55. A summary of the Boeing system is in Konecci, "Bioastronautics Review—1963," 10. The full proposal is R. H. Lowry and Eugene B. Konecci, "An Operating Five-Man, 30-Day Life Support system," presented to the 14th International Astronautical Congress, September 26–October 1, 1963, Folder "Biotechnology & Human Res. Advis. Cte.," Box 1, Series 36: Life Sciences Directorate, 1963–67, RG255, NARA (acc. no. 255-93-022).

56. For the full story, see Lowry and Konecci, "An Operating Five-Man, 30-Day Life Support System."

57. Langley Research Center Life Support Studies"; Harold Klein, "Foreword," in *The Closed Life-Support System*, iii.

58. North American Aviation, Inc., "Manned Mars Landing and Return Mission Study: Final Report," April 1964, Box 1, Series 36: Life Sciences Directorate, 1963–67, RG255, NARA (acc. no. 255-93-022).

59. Drake et al., "Study of Life-Support Systems," 20.

60. Dale W. Jenkins, Office of Space Science and Applications, Chairman, "Summary Report of the Biological Working Panel Manned Space Station Committee," October 11, 1966, Folder "NASA Manned Space Station, Oct 11, 1966," Box 2, Series 36: Life Sciences Directorate, 1963–67, RG255, NARA (acc. no. 255-93-022), section 2.2.4.1; Wilkins, "Man, His Environment and Microbiological Problems," 10.

61. Jenkins, "Summary Report," section 2.2.4.1.

62. Memo from John Billingham, September 28, 1966, Folder "NASA Manned Space Station, Oct 11 1966," Box 2, Series 36: Life Sciences Directorate, 1963–67, RG 255, NARA (acc. no. 255-93-022).

63. Memorandum from the Consultant for Life Sciences to the Administrator to the Associate Administrator, October 4, 1963, Document III-10, in Logsdon et al., *Exploring the Unknown*, 6:346.

64. To many at Oak Ridge, the senior ecologist Stanley Auerbach said, ecologists were seen as "little more than butterfly chasers." Bocking, *Ecologists and Environmental Politics*, 72, 79, 82.

65. Jenkins, "Summary Report," section 2.2.4.1. The literature on biologists' struggle to define their own field and their relationship to physics and the law-centered view of science is huge. See, e.g., Stadler, "Models, the Cell, and the Reformations"; Ankeny, "Wormy Logic," 46–58; Creager, *The Life of a Virus*. On ecology, see Lawton, "Are There General Laws in Ecology?"

66. Hector the Rat was headlined by Eugene Konecci in his talk to the International Astronautics Federation in September 1963. Konecci, "Bioastronautics Review—1963"; Turkina, *Soviet Space Dogs*; Wellerstein, "Remembering Laika."

67. Jenkins, "Summary Report," section 2.2.4.1, 7.

68. John Billingham to Charles Donlan, October 21, 1966," Folder "NASA Manned Space Station, Oct 11, 1966," Box 2, Series 36: Life Sciences Directorate, 1963–67, RG255, NARA (acc. no. 255-93-022); John Naugle to the NASA Deputy Administrator, March 24, 1977, Document III-21, in Logsdon et al., *Exploring the Unknown*, 6:398; Gitelson, "Biological Life-Support Systems," 168.

69. T. Wydeven and E. Smith, "Water-Vapor Electrolysis," in Folder "Life Sciences Accomp.," Box 2, Series 36: Life Sciences Directorate, 1963–67, RG255, NARA (acc. no. 255-93-022). Harold Klein wrote out the reaction in his notes of a meeting: "$2CH_4 + 2CO_2 \rightarrow CH_3OOH$." Handwritten notes ~Nov 1964, Folder "Biotech & Human Res. Comm. 9–11–64," Box 1, Series 36: Life Sciences Directorate, 1963–67, RG255, NARA (acc. no. 255-93-022).

70. Cloud, "Artificial Gills," 72. It was a marker of how seriously Klein took Schnitzer's enquiries that he had all the correspondence translated. Klein to Schnitzer, October 1965; Schnitzer, "Hygiene Work Circle," July 7, 1965; Schnitzer to Klein, October 25, 1965, Folder "Dr. Schnitzer Material on Long Term Life Support Prob.," Box 2, Series 36: Life Sciences Directorate, 1963–67, RG255, NARA (acc. no. 255-93-022).

71. Copy of "Bioscience in the Manned Orbital Space Station" provided to Harold Klein, August 19, 1966, Folder "NASA Manned Space Station, Oct 11, 1966," Box 2, Series 36: Life Sciences Directorate, 1963–67, RG255, NARA (acc. no. 255-93-022); "Draft—Report of Long-Term Flight Panel," September 19, 1966, attached to Memo from John Billingham, September 28, 1966.

72. Memo from John Billingham, September 28, 1966.

73. Space Station Requirements Steering Committee, *Needs and Requirement for a Space Station*, 14. Compton and Benson, *Living and Working in Space*, put NASA's best face forward.

74. Pearson and Grana, *Preliminary Results*, iii, 547. See also Lane et al., *Isolation*, 8–9.

Chapter 2: The Algatron versus the Fecal Bag

1. Armstrong et al., *First on the Moon*; Godwin, *Apollo 9*, 112. Using a sketch of a day in the life on a Red Planet mission from Robert Zubrin, Pat Duggins likewise notes the time set aside for the daily briefing but does not mention the daily evacuation. See Duggins, *Trailblazing Mars*, 194–96.

2. Brooks et al., *Chariots of Apollo*, 268. The food is regularly mentioned, as are several bouts of nausea; Murphy, *Mars*, 62, 91. Identical is Pogue, *How Do You Go to the Bathroom in Space?* The title of Pogue's book promises a frank assessment of the topic, though in fact only 3 of more than 250 questions (nos. 79, 80, and 81) concern the toilet. Better is Mullane, *Do Your Ears Pop in Space?*, 118–22.

3. Clarke, *Promise of Space*, 116. Clarke intrigued his readers by saying that "the ultimate solution to this problem is given in the next chapter," but one searches in vain for a clear description.

4. Peake, *Ask An Astronaut*, 3.

5. Schweickart, "There Ain't No Graceful Way," 117.

6. "We have checklists for everything we did, even how to go to the bathroom there's a checklist," said Apollo 12 command module pilot Dick Gordon. Muir-Harmony, *Apollo to the Moon*, 181. The National Air and Space Museum's display object "Urine Collection and Transfer Assembly, Apollo 11," is number 23 in the collection.

7. Wolfe, *The Right Stuff*, 250. The story and conclusion are basically repeated in Degroot's *Dark Side of the Moon*, 139. Likewise, there is but one paragraph in more than seven hundred pages of Burrow's *This New Ocean*, 359–60.

8. Kelly, *Endurance*, 25; Compton and Benson, *Living and Working in Space*, 131; Chaikin, *A Man on the Moon*, 485; Mullane, *Riding Rockets*, 76.

9. *The Martian*, dir. Ridley Scott, written by Drew Goddard, Twentieth Century Fox, 2015.

10. George, *The Big Necessity*; Simons, "Waste Not, Want Not," 73–98. Notably, Reid, *Paris Sewers and Sewermen*, 3, has shaped our thinking on the social and technological choices of waste management.

11. See Strasser, *Waste and Want*; Rathje and Murphy, *Rubbish!*; Curtis, *Don't Look, Don't Touch, Don't Eat*; Douglas, *Purity and Danger*, 2.

12. Laporte, *History of Shit*. Historian Rose George argued that looking at what people consider shit reveals more about them than does looking at what they display and celebrate. George, *Big Necessity*. See also Thompson, *Rubbish Theory*.

13. Laporte, *History of Shit*, 4, 46; for the edict see pp. 3–7. Hanley, *Everyday Things in Premodern Japan*, 111–12.

14. From M. A. Chevallier, "Sur les urines, les moyens de les recueillir et de les utilizer," *Annales d'hygiène publique et de médicine légale* (January 1852), 68, quoted in Laporte, *History of Shit*, 121; Reid, *Paris Sewers and Sewermen*, 68–70.

15. Duggins, *Trailblazing Mars*, 29; Portree, *Humans to Mars*, 29; Logsdon, "Project Apollo," 421. On the turbulent 1960s, see Matusow, *Unravelling of America*; Sugrue, *Origins of the Urban Crisis*.

16. Pitts, *Human Factor*, 114; Compton and Benson, *Living and Working in Space*, 141. Eating food that produced minimal fecal waste is the only mention of the odious business in *Space* (505), James A. Michener's rival novel to Tom Wolfe's *The Right Stuff*. Godwin, *Apollo 9*, 90. Astronauts' feces from orbital missions are evidently stored in freezers away from the public eye; the room is the NASA equivalent of Narnia, science popularizer Mary Roach was told. Roach, *Packing for Mars*, 275n.

17. Above all, see the fantastic chapter "The Roads Not Taken: Alternative Social and Technical Approaches to Housework" in Cowan, *More Work for Mother*. See also Kevles and Geison, "Experimental Life Sciences," 98; Malone, "Skulking Way of War," 41–52; Perrin, *Giving Up the Gun*; Reynolds and Bernstein, "Edison and 'The Chair,'" 19–28; Volti, "Why Internal Combustion?," 42–47; Kirsch, *Electric Vehicle Company*.

18. MacKenzie, *Knowing Machines*, 7. Looking at failure or just the roads not taken is one cure for the too-easy assumption that successful technologies were somehow inevitable or natural: McCray, *The Visioneers*, 19. Eye-opening was Ard, "Garbage in the Garden State," 57–66. The still-standard starting points are Thompson, *Making of the English Working Class*, rescuing the poor stockinger; and Bailyn, "The Historiography of the Losers," rescuing the defeated Loyalist.

19. Irwin and Emerson, *To Rule the Night*, 43; Pogue, *How Do You Go to the Bathroom in Space?*, 73; Roach, *Packing for Mars*, 273, 275; Godwin, *Apollo 8*, 196. From the evidence amassed by Freidel, *Zipper*, 204–24; and Tone, "Making Room for Rubbers," 51–76, technologies become a "darn thing," always breaking at the most laughably inopportune moments.

20. Roach, *Packing for Mars*, 299. According to James and Alcestis Oberg, the record was set by Wally Shirra piloting the Gemini 6A craft, which rendezvoused with Gemini 7, Oberg and Oberg, *Pioneering Space*, 178, but this is unlikely, as Shirra's flight only lasted twenty-six hours. Data for the duration of American and Soviet manned missions between 1961 and 1967 from table 3, Science Policy Research Division, *Review of the Soviet Space Program*, 15.

21. A tip of the hat to Mary Roach for uncovering this amazing fact. Roach, *Packing for Mars*, 272.

22. Kelly, *Endurance*, 19.

23. Much of this section was previously published as David P. D. Munns and Kärin Nickelsen, “To Live among the Stars: Artificial Environments in the Early Space Age,” *History and Technology* 33 (2018): 272–99. Oswald and Golueke, “Environmental Control Studies,” 183. See Oswald et al., “Closed Ecological Systems,” 45; Golueke et al., “A Study of Fundamental Factors,” 1.

24. The Living Machine in the Findhorn ecovillage, for example, uses a series of connected barrels with plants and algae as a sewage system.

25. Benemann, “Professor William J. Oswald,” 97–98.

26. Report on the Richmond Field Station to dean John R. Whinnery, April 30, 1962, CU149, box 79, folder 11, Archives, Bancroft Library, University of California, Berkeley; Golueke and Oswald, “Role of Plants in Closed Systems,” 389, 393.

27. Oswald et al., “Closed Ecological Systems,” 31, table 1, 30. This type of experimental system paralleled the earlier phytotrons with their extensive, high-technology equipment to standardize and control all environmental parameters. See Munns, *Engineering the Environment*; Shelef et al., “An Improved Algatron Reactor,” 1.

28. See “Prelude: A World of Trons,” in Munns, *Engineering the Environment*, xix.

29. Panel for the Plant Sciences, *Plant Sciences Now*, iii. Quote from Harvey Brooks’s cover letter submitting the report to Seitz at the National Academy of Sciences.

30. Waisel et al., *Plant Roots*, 299–300; Finér et. al, “The Joensuu Dasotrons,” 137–49; Inoue et al., “The ‘Assimitron,’” 165–71.

31. Biotron Conference, December 10–12, 1959, Biotron Papers, Series 06/80, Box 1, Folder “Biotron Conference,” Archives, University of Wisconsin–Madison, 35; Munns, *Engineering the Environment*; Kirk, *Counterculture Green*, 15.

32. Nickelsen, *Explaining Photosynthesis.*

33. According to Sabanas, the female ring-tailed monkey survived the month enclosure without stress, even gaining a pound in weight, as she had a tendency to “overeat.” Sabanas, *Closed Ecological Life-Support Unit.*

34. Golueke and Oswald, “Closing an Ecological System,” 525, 527.

35. Wharton et al., “Algae in Space,” 490–91, 497.

36. Description taken from Oswald et al., “Closed Ecological Systems,” 23–46; Golueke and Oswald, “The Algatron,” 3–9.

37. Nickelsen, “The Organism Strikes Back,” 1–22; Benemann, “Professor William J. Oswald,” 97.

38. Nickelsen, and Govindjee, *Maximum Quantum Yield Controversy*; Nickelsen, “Construction of a Scientific Model,” 73–86; Nickelsen, *Explaining Photosynthesis*; Zallen, “The ‘Light’ Organism for the Job,” 269–79; Nickelsen, “The Organism Strikes Back.” See also Nickelsen, “Physicochemical Biology and Knowledge Transfer.”

39. One of the prominent spokespersons of this position was the conservationist Henry Fairfield Osborn Jr., who wrote enormously popular books on these issues, such as *The Plundered Planet* (1948) and *The Limits of the Earth* (1953). Osborn saw a clear connection between feeding the people of the Earth and fighting communism, and the mass culturing of algae was among his suggested solutions.

40. Belasco, "Algae Burgers for a Hungry World?," 621.

41. Myers, "Basic Remarks," 407–11; Dennis Hevesi, "Jack Myers, 93, Editor Who Geared Science to Children," *New York Times*, January 6, 2007, A16. As Hevesi pointed out, Myers edited the magazine *Highlights*, a general interest science magazine for children with a circulation of about two million.

42. Myers, "Combined Photosynthetic Regenerative Systems," 283. See, e.g., the references in Müntz, "Die Massenkultur von Kleinalgen," S. 311–350.

43. Belasco, *Meals to Come*, 187; Also Belasco, "Algae Burgers for a Hungry World?," 608–634. In 1953 the first experiences with the mass culturing of algae were collected in *Algal Culture*, a volume edited by John S. Burlew from the Geophysical Laboratory of the Carnegie Institution.

44. French, "Photosynthesis," 29.

45. For discussions of what engineers consider "elegant" solutions see, Reid, *Paris Sewers and Sewermen*, 53. Also Hecht, *Radiance of France*, ch. 1; Oswald and Golueke, "Man in Space," 456. For a general discussion of the important and widespread notion of cycle of life, see Ackert, *Sergei Vinogradskii.*

46. Oswald and Golueke, "Environmental Control Studies," 183; Oswald et al., "Closed Ecological Systems," 28.

47. Oswald and Golueke, "Environmental Control Studies," 184; Oswald et al., "Closed Ecological Systems," 24.

48. The theme of human control over space systems is highlighted in Mindell, *Digital Apollo*. In one of their last publications, several of the leaders of the Soviet facility stressed that human control over the environment was given to the occupants. See Gitelson et al., *Manmade Closed Ecological Systems*, 244.

49. Oswald et al., "Closed Ecological Systems," 27; Oswald and Golueke, "Man in Space," 459.

50. For the emphasis on ecological thinking, see Anker, "Ecological Colonization of Space," 246; Poole, *Earthrise*, 2; Höhler, *Spaceship Earth.*

51. Quote attributed to William Oswald in Bewicke and Potter, *Chlorella*, 8.

52. Salisbury et al., "BIOS-3."

53. See, e.g., *Conference on Nutrition in Space*; *The Closed Life-Support System*; *Bioregenerative Systems.*

54. Charles Golueke was the panel discussant for Ott's session. Ott, "Waste Management for Closed Environments," 97–102. There was also the major conference held at NASA's Ames Research Center. *The Closed Life-Support System*, 77; Ott, "Waste Management for Closed Environments," 97, 102.

55. Our thanks to Leah Aronowsky for introducing us to her early work on the Recyclostat, "NASA and the Dream." Krauss, "Discussion," 294. Krauss continued to work on this device up to 1973, funded with NASA grants, see, e.g., his final report, "A Study of Psychophysiology."

56. Krauss, "Discussion," 289. Krauss claimed at a conference in 1966 that his was "the only group who has a recycling system in operation" with the Recyclostat, perhaps unfairly dismissing the Algatron, which he certainly knew about. See the discussion at the end of Krauss, "Physiology and Biochemistry of Algae," 108. The competition to get one's instruments into space was as fierce as that between astronauts.

57. Teller, "Water Generation in Space," 176.

58. Oswald et al., "Closed Ecological Systems," figure 9, p. 43.

59. Oswald et al., "Closed Ecological Systems," 41, 32.

60. Science Policy Research Division, *Review of the Soviet Space Program*, 46; Shelef et al., "An Improved Algatron Reactor," 1.

61. Oswald, "Coming Industry of Controlled Photosynthesis," 235.

Chapter 3: The People's "Planetship"

Epigraph: Gitelson et al., *Manmade Closed Ecological Systems*, 234.

1. See Gaddis, *The United States and the Origins*; Westad, *Global Cold War*; Stephanson, "Cold War Degree Zero," 19–49.

2. Dickson, *Sputnik*; Siddiqi, *Sputnik and the Soviet Space Challenge*; Brzezinski, *Red Moon Rising*, ch. 8.

3. Siddiqi, "Imagine the Cosmos," 79–116.

4. Adas, *Dominance by Design*.

5. On Soviet nuclear power, see Josephson, *Red Atom*.

6. As Sonja Schmid notes, the international exchange of information about the peaceful uses of the atom began in 1955. Schmid, *Producing Power*, 13.

7. Harford, *Korolev*, 161.

8. Widely accessible in the English-speaking world has been Valentin Lebedev's memoir of his long-term stay inside the Salyut space station in 1982. Lebedev, *Diary of a Cosmonaut*. Another valuable source was Bozhko and Gorodinskaya, *A Year in a "Starship,"* ch. 20. We gratefully thank Lloyd Ackert for his invaluable translations of Bozhko's book.

9. Siddiqi, *Sputnik and the Soviet Space Challenge*, 208.

10. For example, see Sisakian, "Contribution of the U.S.S.R.," 23–24.

11. Andrews, *Red Cosmos*; Siddiqi, "Imagine the Cosmos," 80–81; Kojevnikov, "Cultural Spaces of the Soviet Cosmos," 16–21; Siddiqi, *Red Rockets' Glare*.

12. Recent historical work has explored how the Soviet space program opens significant windows into the relationship between science and the state in the post-Stalinist Soviet Union. See, e.g., Andrews and Siddiqi, *Into the Cosmos*, 3–4; Kojevnikov, "Cultural Spaces of the Soviet Cosmos," 22–27.

13. Noted by the Office of Technology Assessment, *Salyut: Soviet Steps towards Permanent Human Presence in Space*, December 1983, Box 5, Series: R&D Projects—Soviet Space Study, RG255, NARA (acc. no. 255-09-005); Oberg and Oberg, *Pioneering Space*, 280. The U.S. space shuttle fleet would amass a total of 1,322 days in space, which across seven crewmembers equals 9,254 man-days in space.

14. Paine, "A Timeline for Martian Pioneers," 13.

15. Lebedev, *Diary of a Cosmonaut*, vii.

16. Kirensky et al., "Theoretical and Experimental Decisions," 75; Gitelson, "Biological Life-Support Systems," 167.

17. Summary quote of the Fourth Paris Lecture on Space Medicine, reported to the Foreign Broadcast Information Service. In Joint Publications Research Service, *USSR Report: Space* JPRS-USP-86-005, September 12, 1986, 140, Box 6, R&D Projects—Soviet Space Study, RG255, NARA (acc. no. 255-09-005).

18. Introductions to the BIOS project in English include Salisbury et al., "BIOS-3"; Gitelson et al., *Manmade Closed Ecological Systems*, chapter 9. For a review of space based plant growth experiments, notably the Soviet ones, see Zabel et al., "Review and Analysis," 1–16. It should be stressed that we are relying on a collection of published sources translated into English from Russian originals. These originally came through a specifically commissioned service as part of the American effort to learn about the other side during the Cold War, while today they are part of the extensive NASA database ntrs.nasa.gov as well as the NARA files in RG255 related to life science at NASA. There are, furthermore, extremely valuable accounts from former protagonists of the Soviet program. Some of them provide substantial detail of the actual development of the life support facilities and the scientific goals that were pursued at the time. Based on these sources, we can at least begin to establish some aspects of the work of Soviet life scientists and environmental engineers toward long-term habitability in space.

19. Turkina, *Soviet Space Dogs*. Soviet space dogs became overnight celebrities, see Nelson, "Cold War Celebrity," 133–55.

20. Sisakian, "Contribution of the U.S.S.R.," 25; Siddiqi, *Challenge to Apollo*, 265.

21. Sisakian, "Contribution of the U.S.S.R.," 29.

22. Of course, the quip was principally intended to ridicule American faith from the point of view of Soviet atheism. See e.g. Siddiqi, "Imagine the Cosmos," 80. In contrast, American astronaut (Apollo 15) Jim Irwin had a deeply spiritual experience in orbit: Irwin and Emerson, *To Rule the Night*, 1973.

23. Sharpe, *Living in Space*, 61–62; Siddiqi, *Challenge to Apollo*, 207; Shelton, *Soviet Space Exploration*, 112.

24. Harford, *Korolev*.

25. Bjurstedt, *Proceedings of the First International Symposium*; Sisakian, "Contribution of the U.S.S.R.," 32.

26. Bozhko and Gorodinskaya, A *Year in a "Starship,"* ch. 20.

27. Shepelev, "Some Aspects of Human Ecology," 166–67.

28. Batov et al., "Dense Continuous Cultivation of Chlorella," 652. See also Meleshko and Krasotchenko, "Conditions of Carbon Nutrition of Chlorella in Intensive Cultures"; and Tsvetkova, Shaydarov, and Abramova, "Special Features of Plant Feeding under Conditions of Aeroponic Cultivation for a Closed System," 637.

29. Salisbury et al., "BIOS-3," 576; Gitelson et al., *Manmade Closed Ecological Systems*, 216–17. It is remarkable that Soviet life scientists felt the need to prove that humans and algae were able to exchange gases; it is even more remarkable that Americans still felt compelled to do so in the 1990s when preparing for the International Space Station, as we shall see in chapter 4.

30. Gitelson et al., *Manmade Closed Ecological Systems*, 218.

31. Gitelson et al., *Manmade Closed Ecological Systems*, 214–15.

32. Shepelev, "Some Aspects of Human Ecology," 170, 167.

33. Levashov, "New Aspects of Personal Hygiene," 164.

34. Siddiqi, *Sputnik and the Soviet Space Challenge*, 334–37, 745.

35. Salisbury et al., "BIOS-3," 576, 577.

36. During the 1950s the Soviet Union had already built the world's largest phytotron. See Munns, *Engineering the Environment*.

37. Bozhko and Gorodinskaya, A *Year in a "Starship,"* ch. 13.

38. Bozhko and Gorodinskaya, A *Year in a "Starship,"* ch. 15.

39. Shepelev, "Some Aspects of Human Ecology," 173–74.

40. Bozhko and Gorodinskaya, A *Year in a "Starship,"* chs. 1, 3.

41. Gorbov and Novikov, "Experimental Psychological Investigation," 15, 19. The device was first described in William Ross Ashby's book, *Design for a Brain: The Origin of Adaptive Behaviour* (1952). See Pickering, *Cybernetic Brain*, ch. 4. Interestingly, Gorbov's experimental setup is cited in recent management circles to understand group dynamics. See Voskoboynikov, *Psychology of Effective Management*, 65.

42. Bozhko and Gorodinskaya, A *Year in a "Starship,"* ch. 20.

43. Bozhko and Gorodinskaya, A *Year in a "Starship,"* ch. 6.

44. Bozhko and Gorodinskaya, A *Year in a "Starship,"* chs. 30, 9. Gitelson et al., *Manmade Closed Ecological Systems*, 267.

45. Bozhko and Gorodinskaya, A *Year in a "Starship,"* ch. 10.

46. http://scienceillustrated.com.au/blog/in-the-mag/dreaming-of-mars-part-1.

47. Gerovitch, "New Soviet Man inside Machine," 135–57.

48. Mashinskiy and Nechitaylo, "Birth of Space Plant Growing," 3, 6; Lebedev, *Diary of a Cosmonaut*, 155, 210–11; Oberg, "Russians to Mars?," 74.

49. Gitelson et al., *Manmade Closed Ecological Systems*, 38, 245, 247–48, 308.

50. Gitelson et al., *Manmade Closed Ecological Systems*, 234.

51. Lebedev, *Diary of a Cosmonaut*, 37.

52. Gitelson et al., "Long-Term Experiments," 67; I. A. Terskov, I. I. Gitel'zon, B. G Kovrov, G. M. Lisovsky, Yu. N. Okladnikov, F. Ya. Sid'ko, V. N. Belyanin, and M. P. Shilenko, "Experiment with the Incorporation of Vegetable Plants in a Semiclosed Life Support System,' *Space Biology and Medicine* 8, no. 3 (1974): 53–58, trans. U.S. Joint Publications Research Service (JPRS 62553), Box 1, R&D Projects—Soviet Space Study, 1960–90, RG 255, NARA (acc. no. 255-09-005).

53. Gitelson et al., "Long-Term Experiments," 69.

54. "Quails in Space."

55. Gitelson et al., *Manmade Closed Ecological Systems*, 29. See also Mashinskiy and Nechitaylo, "Birth of Space Plant Growing," 3.

56. Roll-Hansen, *The Lysenko Effect*; Levins and Lewontin, "The Problem of Lysenkoism," in *The Dialectical Biologist*, 163–96.

57. Rasmussen, *Gene Jockeys*, 25; Munns, *Engineering the Environment*, 17–18. In the reformation of American science education in the wake of Sputnik, genetics assumed an inflated role in the new curriculum because it was anti-communist. See Rudolph, *Scientists in the Classroom*, 52.

58. On the political uses of Lysenko for American science, see Wolfe, *Freedom's Laboratory*, ch. 1; Andrews and Siddiqi, *Into the Cosmos*, 3; Siddiqi, *Red Rockets' Glare*, 6.

59. Office of Technology Assessment, *Salyut*, 43.

60. Oberg, "Russians To Mars?," 74; DeMeis, "Mir," 24; *Time*, October 5, 1987.

61. Gitelson et al., "Long-Term Experiments," 66. Astronauts going to the ISS have to launch from Baikonur. See, for example, Kelly, *Endurance*.

62. Gitelson et al., *Manmade Closed Ecological Systems*, 220.

Chapter 4: Gardens in Space

1. Bush, *Pieces of the Action*, 8.

2. Quoted in Egan, *Barry Commoner*, 140. The decisive warriors of the 1960s included Jacobs, *Death and Life*; Carson, *Silent Spring*; Nader, *Unsafe at Any Speed*; Commoner, *Closing Circle*; Bernstein and Woodward, *All the President's Men*; Vidal, "State of the Union," 265–85. Criticism of the military-industrial complex became a focus of the New Left and the environmental movement; see Gottlieb, *Forcing the Spring*, 138.

3. Parker, "Managing Space," 28–29; Nelkin, *University and Military Research*; Wisnioski, *Engineers for Change*; Mody, "Santa Barbara Physicists," 70–107; Tibbe, *No Requiem for the Space Age*, 135–36; Mailer, *Of a Fire on the Moon*, 125. Underappreciated at the time, the image of the new frontier of space opened a place for Americans to reconsider their racial assumptions. See Kilgore, *Astrofuturism*.

4. "The Lavatory Thing," in Bizony, *How to Build Your Own Spaceship*, 145.

5. Canby, "Skylab," 441–92. Public Affairs Office, Lyndon B. Johnson Space Center, "Skylab—1973–1974," *NASA Facts* (JSC 08826); Compton and Benson, *Living and Working in Space*, 152, 158.

6. Gorman, "The Sky is Falling," 529–46; "Possible Planning Assumptions and Their Implications for the Long-Range Space Program, 7/2/75," Folder "NASA 1975: General Jan–July," Box 21, Glenn R. Schleede Papers, Gerald R. Ford Library; Conway, *Exploration and Engineering*. See also Smith, *The Space Telescope*. The Voyager probes also supplied the central storyline for the motion picture reboot of the television show *Star Trek*, *Star Trek: The Motion Picture* in 1979. "Background Materials for the Director's Meeting with Dr. Fletcher," July 7, 1975, Folder "NASA 1975: General Jan–July,' Box 21, Glenn R. Schleede Papers, Gerald R. Ford Library, p. 3. A line by novelist Ben Bova aptly summarized NASA's situation in the mid-1970s: "Nobody's against space, but they just don't have space high on their priority list." Bova, *Mars, Inc.*, 78.

7. Paine, "A Timeline for Martian Pioneers," 12.

8. Mody, "Square Scientists and the Excluded Middle," 58–71; Kaiser and McCray, *Groovy Science*. For a contrasting perspective on the transformation of science in the 1970s, see Egan, "Survival Science," 26–39.

9. McCray, *The Visioneers*, 60–66. For an extensive treatment of Dyson's vision of space travel, contrasted with his son's ideas about how to survive on Earth, see Brower, *The Starship and the Canoe*. On Dyson's earlier work, see Kaiser, *Drawing Theories Apart*, ch. 3. For Dyson's own account, see Dyson, *Disturbing the Universe*.

10. In a moment of scientific celebrity, O'Neill described his ambitious space colonization plan to the U.S. House committee on science and technology in July 1975. "Space Colonization and Energy Supply," 12–21. See also Stewart Brand's interview with Gerard O'Neill, *Space Colonies*, 22–30. On Drexler's later career in nanotechnology, see Milburn, *Nanovision*.

11. McCray, *The Visioneers*, 10–13, 69; Heppenheimer, *Colonies in Space*, 26. The sexiness of the astronaut-engineer is at the heart of Tom Wolfe's *The Right Stuff*. Both Kidder, *Soul of a New Machine* and, later, Cringely, *Accidental Empires* stressed the expectations of gender of the high-tech industry in the 1970s and 1980s.

12. Klerkx, *Lost in Space*, 76; McCray, *The Visioneers*, 61. Repeating the easy solution of the 1950s, reliable and easy-to use nuclear reactors power the Mars novels of both Ben Bova and Kim Stanley Robinson: Bova, *Mars*, 25; Robinson, *Red Mars*, 105.

13. Sharpe, *Living in Space.* Sharpe's book grew out of his earlier work with the George C. Marshall Space Flight Center. See Sharpe, "Human Parameter." Drexler, "Canvas of the Night," 43.

14. Robinson, *Red Mars,* 133–34; "The Assignment," *Star Trek: Deep Space Nine,* season 5, episode 5; Weir, *Artemis,* 95. It was an attitude repeated in a science fiction story from 2000 by James Patrick Kelly, whose protagonist agriculture technician worked to keep microorganisms in balance to grow crops: "It's important work," he says, "and duller than accounting." Kelly, "Breakaway, Backdown," 100.

15. On Warshall, see Kirk, *Counterculture Green,* 94–100. Peter Warshall was later in charge of selecting vertebrates for the Biosphere 2. See Nelson, *Pushing Our Limits,* 119.

16. Comments from Lynn Margulis and John Todd, Brand, *Space Colonies,* 35, 49; Trim, "A Quest for Permanence," 142–71.

17. Brower, *The Starship and the Canoe,* 56–57; O'Neill, *2081,* 108; comments from Peter Warshall, Brand, *Space Colonies,* 52.

18. The goal for Heppenheimer was ultimately population control, particularly in the "underdeveloped countries," as he put it. His vision, in other words, not only assumed that resources and energy would be abundant in space, but also combined that with a racial undercurrent. Heppenheimer, *Colonies in Space,* 184, 152, 131–32, and ch. 11. Sex gets more discussion than bathrooms also in Bova, *Mars, Inc.,* 424.

19. Oswald and Golueke, "Man in Space," 456; Wilkins, "Man, His Environment and Microbiological Problems."

20. O'Neill, *2081,* 238, 64; Schweickart, "There Ain't No Graceful Way," 116–19. See "Jacques Cousteau at NASA Headquarters," in Brand, *Space Colonies,* 98–103.

21. Mark, *The Space Station,* 67; Neal, *Spaceflight in the Shuttle Era,* ch. 2, 37, 44, 42.

22. See Klerkx, *Lost in Space,* 15; Entry from June 9, 1977, Carter, *White House Diary,* 63; quoted in Lane et al., *Isolation,* 1.

23. Muenger, *Searching the Horizon*; Messeri, *Placing Outer Space,* 75–76.

24. Wharton et al., "Algae in Space," 488; Gitelson et al., *Manmade Closed Ecological Systems,* 38; Wheeler, "Bioregenerative Life Support and Nutritional Implications," 62.

25. MacElroy and Averner, "Space Ecosynthesis," 1. See Dick and Strick, *Living Universe,* 39.

26. MacElroy and Averner, "Space Ecosynthesis," 15, 13.

27. MacElroy and Averner, "Space Ecosynthesis," 17, 1, 9, 24. See also Gitelson et al., *Manmade Closed Ecological Systems,* 36.

28. Gitelson et al., *Manmade Closed Ecological Systems,* 37. See MacElroy and Averner, "Space Ecosynthesis."

29. Byrd, "The Long Road Starts Here," 28. For Folsome's work, see Folsome and Hanson, "Emergence of Materially-Closed-System Ecology," 269–88.

30. Mason and Carden, *Controlled Ecological Life Support System,* 31, 39.

31. Mason and Carden, *Controlled Ecological Life Support System,* 13–14, 9, 10, 12, 35, 36.

32. Mason and Carden, *Controlled Ecological Life Support System,* 28–29.

33. Shiro Furukawa, "Closed Ecological Life Support System," in *Life Sciences Considerations for Long Duration Manned Space Missions,* volume. 1: Medical Operation, NASA Contract NAS8-3235, KSC-1, R & D Projects—Space Station Program Reports, 1982–1984, Box 2, RG255, NARA (acc. no. 255-09-001), 5–8.

34. Mark, *The Space Station*, 152; excerpt, "Space Station Program Briefing," slide 1, AMES Research Center, February 24, 1984, R&D Projects—Space Station Program Reports, 1982–1984, Box 2, RG255, NARA (acc. no. 255-09-001).

35. Few popular culture references even remain, with Bova, *Mars*, 196, an exception. For the internal political story, see Mark, *The Space Station*, chs. 11–12. Schmitt, "A Millennium Project," 30. In classic colonialist and great power thinking, Schmitt advocated for an American Mars project, asking the audience, "how sad if this adverse trend of history (Soviet commemorations) is established by the 500th Anniversary of the discovery of America by Columbus?" "Space Station Program Briefing."

36. Mark, *The Space Station*, 145, 193–96, Appendix 10, 137; "Space Station Program Briefing"; "Workshop Briefing Charts," NASA Space Station Task Force Concept Development Group, December 5–9, 1983, R&D Projects—Space Station Program Reports, 1982–1984, Box 4, RG255, NARA (acc. no. 255-09-001).

37. "Workshop Briefing Charts"; John B. Hall and Shelby J. Pickett, "Environmental Control and Life Support System (ECLSS) Technology Assessment for Manned Space Stations," Concept Development Workshop, July 11–13, 1983, R&D Projects—Space Station Program Reports, 1982–1984, Box 3, RG255, NARA (acc. no. 255-09-001); Salisbury, "Controlled Environment Life Support Systems (CELSS)."

38. See CDG Study Task #7—Common Module, and Brian O'Leary, "The Case for Long Modules," in "Workshop Briefing Charts."

39. Furukawa, "Closed Ecological Life Support System," 5–10.

40. Ware and Chandler, "Civilian Space Program," 52; Mandell, "Space Station," 157–170.

41. Kelly, *Endurance*, 51.

42. Boeing Aerospace Company, *Space Station/Soviet Space Station Analog: Soviet Literature Report, Space Station Crew System Interface Study* (D180-28182-1), NASA, October 1983, R&D Projects—Soviet Space Study, 1960–90, Box 6, RG255, NARA (acc. no. 255-09-005), 46, 49; Compton and Benson, *Living and Working in Space*; Pogue, *How Do You Go to the Bathroom in Space?*, question 58, 56–57.

43. "Environmental Requirements in Biotron, Year 1990–91," Biotron Papers, Series 06/80, Box 4, file "UW Graduate School—Biological Sciences Review, 1991," Archives, University of Wisconsin—Madison; Tibbitts et al., *Growth of Potatoes for CELSS*, iii, vi–vii.

44. Timmins, "From Space Suits to Space Couture," 194; Mason and Carden, *Controlled Ecological Life Support System*, 9, 10, 12.

45. Furukawa, "Closed Ecological Life Support System," 5–10.

46. Fortunately and excitingly, historical work on the psychology of space travel is taking shape, see Jordan Bimm's review of two recent works by Douglas Vakoch: Bimm, "Psychology of Space Exploration," 812. In addition, Oliver, "Psychological Effects of Isolation."

47. A survey of such research from 1984 is Harrison and Connors, "Psychological and Interpersonal Adaptation," 643–54.

48. Comments by Bruce Pittman at the Space Station Human Factors Research Review: Architecture Panel Discussion, December 3–6, 1985, in *Space Station Human Factors Research Review*, 193.

49. Borisov, "Who Should Be Entrusted," 4. NASA echoed this Soviet research, including looking at Heyerdahl's role in his expeditions. See Connors et al., *Living Aloft*, 164–65. There

is a great project to be done in exploring the parallels between the technical innovation and adventurous expeditions of Heyerdahl reed rafts and NASA's Mercury and Gemini projects. See Heyerdahl, *Ra Expeditions*.

50. Packham, "Lunar-Mars Life Support Test Project," 23; Oberg and Oberg, *Pioneering Space*, 181.

51. Von Puttkamer, "Human Role in Space." An expanded version of his overhead-projector slide presentation that offers a manned Mars mission as a larger goal of any space station is von Puttkamer, "Beyond the Space Station," 171–206.

52. Sharon Matsumura to Trieve Tanner, Chief, Human Factors Office, NASA, October 20, 1983, Box 4, "Ames Research Center Projects & Systems Engineering branch; R&D projects—Soviet Space Study, 1982–1984," RG255, NARA (acc. no. 255-09-001), 2; McMurray v. Phelps, 533 F. Supp. 742.

53. O'Leary, "The Case for Long Modules."

54. H.T. Fisher, Crew Systems, Space Systems Division, Lockheed Missiles and Space Company, "Habitability Sleep Accommodations," Presentation to NASA Space Station Task Force Concept Development Group, December, 5–9, 1983, Box 4, R&D Projects—Space Station Program Reports, 1982–1984, RG255, NARA (acc. no. 255-09-001); Keith Miller and B. J. Bluth, "Habitability Studies at Boeing," in "Workshop Briefing Charts."

55. Bova, *Mars*, 211.

56. Litton, "Issue Coordinator's Comments," 91; Walker and Granjou, "MELiSSA the Minimal Biosphere," 59–69; Lawton, "The Ecotron," 181–94; Kibe et al., "Controlled Ecological Life Support," 117–25.

57. Wheeler, "Bioregenerative Life Support and Nutritional Implications," 50, 46; Knott, "Breadboard Project," 45; Wheeler et al., "System Development and Early Biomass Tests"; Dickey, "Seeding Space," 15–19.

58. Packham, "Lunar-Mars Life Support Test Project," 17.

59. Henninger, "Test Phases and Major Findings," 39.

60. Henninger, "Test Phases and Major Findings," 44.

61. Henninger, "Test Phases and Major Findings," 45.

62. Henninger, "Test Phases and Major Findings," 47.

63. Connolly, "Architecture," 80; Henninger, "Test Phases and Major Findings," 47.

64. Henninger, "Test Phases and Major Findings," 37.

65. Kelly, *Endurance*, 190, 344–45. For more on NASA's Veg experiments, see "Vegetable Production System," Space Station Research Explorer, nasa.gov.

Chapter 5: Escaping Earth in Biosphere 2

Mark Nelson, *Pushing Our Limits: Insights from Biosphere 2* (Tucson: University of Arizona Press, 2018), 35.

1. Reider, *Dreaming the Biosphere*, 66; Peder Anker, Dana Fritz, Linda Leigh, Lisa Ruth Rand, and Shawn Rosenheim, "Biosphere 2: Why an Eccentric Ecological Experiment Still Matters 25 Years Later," http://edgeeffects.net/biosphere-2/; Anker, *From Bauhaus to Ecohouse*; Nelson, "Biotechnology of Space Biospheres," 193; Nelson, *Pushing Our Limits*; Allen, "Historical Overview"; Allen, *Biosphere 2*.

2. "About Us," Institute of Ecotechnics, November 9, 2017, ecotechnics.edu.

3. General background on the Biosphere 2 is in Nelson, *Pushing Our Limits*; Sagan and Margulis, *Biospheres*; Allen, *Biosphere 2*, 13–14, 16; Alling and Nelson, *Life under Glass*, 196; Allen, "Historical Overview of the Biosphere 2 Project." Gathering many of the voices of the project is Reider, *Dreaming the Biosphere*.

4. Höhler, *Spaceship Earth*, 119; Folsome and Hanson, "Emergence of Materially-Closed-System Ecology," 271; Sabine Höhler, "Ecospheres: Model and Laboratory for Earth's Environment," *Technosphere Magazine*, 2018, https://technosphere-magazine.hkw.de/; Kelly, *Out of Control*, 132–33; and Dempster, "Engineering of Biosphere 2," 109–116.

5. Nelson, "Biotechnology of Space Biospheres," 189–90; Gentry and Liptak, *The Glass Ark*, 29, 77; Dempster, "Methods for Measurement," 331–35.

6. Gentry and Liptak, *The Glass Ark*, 69.

7. Nelson and Soffen, "Foreword," viii; Nelson et al., "Biosphere 2," 211–17; Alling and Nelson, *Life under Glass*, ix; Allen, "Brief History," 93–94.

8. Nelson, *Pushing Our Limits*, 34, 35.

9. Höhler, *Spaceship Earth*, 120. The water loop, for example, is described in Nelson et al., "Water Cycle in Closed Ecological Systems," 1404–12; Reider, *Dreaming the Biosphere*, 181; Gentry and Liptak, *The Glass Ark*, 63, 52. Only in such comparisons can the true innovations of the Biosphere 2 complex be apparent.

10. On the critiques of biosphere and Spaceship Earth by the biospherians, see Höhler, *Spaceship Earth*, 120.

11. Höhler, *Spaceship Earth*, 2, 17, 66. Fuller's Spaceship Earth was a widespread notion. See also Allen, *Biosphere 2*, 9.

12. Odum, *Ecology*, 1. See, however, Mark Nelson's long critique to Anker et al., "Biosphere 2"; Nelson and Soffen, "Foreword," viii.

13. Hagen, *An Entangled Bank*; Kingsland, *Evolution of American Ecology*; Coleman, *Big Ecology*; Creager, *Life Atomic*, ch. 10. Creager noted that early proponents of using radioisotopes as tracers often "did not reckon with the biological effects of the radiation they put into their systems" and that they even claimed that "low-level amounts of radiation did not disturb fundamental living processes" (223).

14. Constance, Interview with Ann Lage, 142; Alling and Nelson, *Life under Glass*, 14; Allen, *Biosphere 2*, 116.

15. Allen, *Biosphere 2*, 6; Reider, *Dreaming the Biosphere*, 116.

16. Nelson, like Allen, stresses that the Biosphere 2 had its origin in the ideas of Russian scientist Vladimir Vernadsky, Eugene Odum's conception of system ecology, as well as the work of the Institute of Ecotechnics.

17. Gitelson et al., *Manmade Closed Ecological Systems*, 2; Kirensky et al., "Theoretical and Experimental Decisions," 75; Nelson, *Pushing Our Limits*, 5–6; Alling and Nelson, *Life under Glass*, 196.

18. Allen, *Biosphere 2*, 22.

19. Cooke, "Ecology of Space Travel," 498.

20. Aronowsky, "NASA and the Dream," 371–72; Folsome and Hanson, "Emergence of Materially-Closed-System Ecology," 271; Gitelson et al., *Manmade Closed Ecological Systems*, 367.

21. Most recently, Anker et al., "Biosphere 2." This argument is fully given in Anker, "Ecological Colonization of Space"; and Anker, *From Bauhaus to Ecohouse*. Höhler, "Eco-

spheres"; Anker, "Ecological Colonization of Space," 256; Nelson, *Pushing Our Limits*, 119, xii; Allen, *Me and the Biospheres*, 127. Nelson noted that Schweickart served on the review committee and "became a friend." Allen credited Schweickart with helping the group to not get carried away by smaller issues but stay focused on the "biggest problems." Also Brand, *Space Colonies*, 116–19.

22. Paine, "Biospheres and Solar System Exploration," 1, 10.

23. Reider, *Dreaming the Biosphere*, 96, 100. Unknowingly, they thus replicated experiments by the famed plant physiologist and creator of the first phytotron Frits Went in 1952: Went, "Earhart Plant Research Laboratory," 14–18.

24. Nelson, *Pushing Our Limits*, 213, Alling and Nelson, *Life under Glass*, viii, dedication.

25. Poynter, *Human Experiment*, vii; Allen, *Biosphere 2*, 3; Alling and Nelson, *Life under Glass*,18; Nelson, *Pushing Our Limits*, 100–101.

26. Allen, *Biosphere 2*, 123.

27. Alling and Nelson, *Life under Glass*, 25, 115, 188; Brin, "Ice Pilot," 53–66, 55.

28. Marc Cooper, "The Profits of Doom: The Biospherians Lure Scientists to a High-Priced Feast under Glass," *Village Voice*, June 19, 1991.

29. Gentry and Liptak, *The Glass Ark*, 69, 70, 73; Alling and Nelson, *Life under Glass*, 71; Poynter, *Human Experiment*, 189. Nelson writes that 25 percent of the crew time was spent in farming. Nelson, *Pushing Our Limits*, 192, 153, 80.

30. The group was Allen, Taber, Gaie, Laser, and Poynter. Poynter, *Human Experiment*, 124. Quoted in Marc Cooper, "Negative Biofeedback," *Phoenix New Times*, June 19, 1991, phoenixnewtimes.com.

31. Keller, "Nature, Nurture, and the Human Genome," 290; Kermode, *Silent Running*, 29, 14, 18, 21; *Silent Running*, dir. Douglas Trumbull. See also Milburn, "Nanotechnology," 109–29; and Milburn, "Nanowarriors," 77–103.

32. Alling and Nelson, *Life under Glass*, 6.

33. Nelson, *Pushing Our Limits*, 188, 194, 199.

34. It was a "cockroach war" every night in the kitchen. See Nelson, *Pushing Our Limits*, 78. Broecker and Kunzig, *Fixing Climate*, 201; Gitelson et al., "Long-Term Experiments," 67.

35. Nelson, "Biotechnology of Space Biospheres," 190; Broecker and Kunzig, *Fixing Climate*, 200; Nelson, *Pushing Our Limits*, 175–76.

36. Reider, *Dreaming the Biosphere*, 243. Reider evidently did not know or appreciate that Osmond came from phytotronic stock via the Australian phytotron under Frankel and Evans. See Lloyd Evans, "Memoirs of a Meandering Biologist," June 2005 National Library of Australia, MS9885, p. 44.

37. Hagen, "Teaching Ecology," 709; Marino and Odum, "Biosphere 2," 10.

38. Nelson, *Pushing Our Limits*, 4.

39. Marino and Odum, "Biosphere 2," 4, 12; Poynter, *Human Experiment*, 203, 341–42; Reider, *Dreaming the Biosphere*, 71.

Conclusion

1. Logan Hill, "Welcome to Mars," *Wired*, November 2016, 135.

2. Gitelson et al., *Manmade Closed Ecological Systems*, 25; Hinghofer-Szalkay and Moore, "Some Comments," 542–61.

3. A survey of the wide range of closed environmental ideas and systems is Yensen and Biel, "Criticality Concept," 29–71.

4. See Taub, "Closed Ecological Systems," 154. It is worth highlighting again that NASA's Technical Reports Server is a treasure trove of information on most topics of space research and development. https://ntrs.nasa.gov.

5. Eric Roston, "Why Elon Musk's Mars Vision Needs 'Some Real Imagination,'" *Bloomberg* October 17, 2016, bloomberg.com; Robinson, *Aurora*, 13.

6. Mashinskiy and Nechitaylo, "Birth of Space Plant Growing," 2, 7, 4; Kidger, "Salyut 6 Mission Report," 53; Bozhko and Gorodinskaya, *A Year in a "Starship,"* ch. 13.

7. See Hawaii Space Exploration Analog and Simulation website, hi-seas.org. On Masdar City, see Günel, *Spaceship in the Desert*. In the imaginings of Masdar City and in its initial construction, "climate change is a management problem that experts may resolve, rather than an ethical and moral problem" (11).

8. Zubrin, *Mars on Earth*, 214; "The Mars 500 Isolation Experiment," 306–8; Mark Garcia, "Astronauts Switch Roles from Scientists to Plumbers," NASA, April 24, 2018, https://blogs.nasa.gov/. In short, astronauts now "go to work" in space exactly as NASA envisioned in the 1970s. See Neal, *Spaceflight in the Shuttle Era*, 78.

9. Gitelson et al., *Manmade Closed Ecological Systems*, 2, 23; Oswald et al., "Closed Ecological Systems," 23; Nelson, *Pushing Our Limits*, 224; Friedel, "History, Sustainability, and Choice," 219–25.

10. Gitelson et al., *Manmade Closed Ecological Systems*, 368.

BIBLIOGRAPHY

Archival and Primary Sources

Allen, John. "Historical Overview of the Biosphere 2 Project." Washington, D.C.: NASA, N91-13844, 1990. Available at NASA Technical Reports Server, http://ntrs.nasa.gov.

Belasco, Norman, and Perry, Donald M. "Waste Management and Personal Hygiene for Extended Spacecraft Missions." Presentation to the American Industrial Hygiene Conference, Philadelphia. NASA-TM-X-57096, 1964. Available at NASA Technical Reports Server, http://ntrs.nasa.gov.

Biology Division Papers. Archives. California Institute of Technology.

Bioregenerative Systems. NASA SP-165, 1968. Available at NASA Technical Reports Server, http://ntrs.nasa.gov.

Biotron Papers. University of Wisconsin–Madison Archives.

Borisov, O. "Who Should Be Entrusted with an Interplanetary Spacecraft?" *Pravda*, April 3, 1974, p. 4. NASA Technical Translation TTF-15,644, 1974. Available at NASA Technical Reports Server, http://ntrs.nasa.gov.

Brooks, Courtney G., and Ivan D. Ertel. *The Apollo Spacecraft—A Chronology.* Volume 3, *October 1, 1964—January 20, 1966.* NASA SP-4009, 1973. Available at NASA Technical Reports Server, http://ntrs.nasa.gov.

The Closed Life-Support System. Ames Research Center, NASA SP-134, 1967. Available at NASA Technical Reports Server, http://ntrs.nasa.gov.

Conference on Nutrition in Space and Related Waste Problems. NASA SP-70, 1964. Available at NASA Technical Reports Server, http://ntrs.nasa.gov.

Connors, Mary M., Albert A. Harrison, and Faren R. Atkins. *Living Aloft: Human Requirements for Extended Spaceflight.* Washington, D.C.: NASA Scientific and Technical Information Branch, 1985. Available at NASA Technical Reports Server, http://ntrs.nasa.gov.

Constance, Lincoln. Interview with Ann Lage. University of California at Berkeley, University History Series, 1986.

Drake, G. L., C. D. King, W. A. Johnson, and E. A. Zuraw. "Study of Life-Support Systems for Space Missions Exceeding One Year in Duration." In Ames Research Center, *The Closed Life-Support System*, 1–74. NASA SP-134, 1967. Available at NASA Technical Reports Server, http://ntrs.nasa.gov.

Evans, Lloyd. "Memoirs of a Meandering Biologist." June 2005. National Library of Australia, MS9885.

Golueke, Clarence G., William J. Oswald., and H. K. Gee. "A Study of Fundamental Factors Pertinent to Microbiological Waste Conversion in Control of Isolated Environments." March 31, 1965. Project No. 8659 under Contract No. AF 19(628)-2462 for the Air Force Cambridge Research Laboratories, USAF.

Haeselin, Dave. "Earth First, Then Mars: An Interview with Kim Stanley Robinson." *Public Books*. June 15, 2016. Available at http://davidhaeselin.com/.

Kramer, Paul Papers. Duke University Archives.

Krauss, Robert W. "A Study of Psychophysiology in controlled Environments." Department of Botany, University of Maryland, NASA Technical Report No. 1024. Available at NASA Technical Reports Server, http://ntrs.nasa.gov.

Ley, Willy. Collection. National Air and Space Museum.

Logsdon, John M., Stephen J. Garber, Roger D. Launius, and Ray A. Williamson. *Exploring the Unknown: Selected Documents in the History of the U.S. Civil Space Program.* Vol. 6, *Space and Earth Science.* Washington, D.C.: NASA, 2004. Available at NASA Technical Reports Server, http://ntrs.nasa.gov.

Logsdon, John M., Linda J. Lear, Jannelle Warren-Findley, Ray A. Williamson, and Dwayne A. Day, eds. *Exploring the Unknown: Selected Documents in the History of the U.S. Civil Space Program.* Vol. 1, *Organizing for Exploration.* Washington, D.C.: NASA, 1995. Available at NASA Technical Reports Server, http://ntrs.nasa.gov.

MacElroy, R. D., and M. M. Averner. *Space Ecosynthesis: An Approach to the Design of Closed Ecosystems for Use in Space.* NASA Technical Memorandum TM 78491, 1978.

Mason, Robert M., and John L. Carden, eds. *Controlled Ecological Life Support System: Research and Development Guidelines.* Washington, D.C.: NASA Conference Publication 2232, 1982. Available at NASA Technical Reports Server, http://ntrs.nasa.gov.

National Aeronautics and Space Administration. *The Challenge of Space Exploration: A Technical Introduction to Space.* Washington, D.C.: NASA, 1959. Available at NASA Technical Reports Server, http://ntrs.nasa.gov.

Ohya, Haruhiko, Oshima, Tairo, and Nitta, Keiji. "Survey of CELSS Concepts and Preliminary Research in Japan." In Robert D. MacElroy, David T. Smernoff, and Harold P. Klein, *Controlled Ecological Life Support System: Life Support Systems in Space Travel*, 10–16. NASA Conference Publication 2378, 1985. Available at NASA Technical Reports Server, http://ntrs.nasa.gov.

Pearson, Albin O., and David C. Grana, comp. *Preliminary Results from an Operational*

90-Day Manned Test of a Regenerative Life Support System. NASA SP-261, 1971. Available at NASA Technical Reports Server, http://ntrs.nasa.gov.

Records of NASA. RG 255. National Archives and Records Administration, San Bruno, CA.

Records of the Office of the Chancellor, UC–Berkeley. CU149. Bancroft Library, University of California, Berkeley.

Schleede, Glenn R. Papers. Gerald R. Ford Library. Ann Arbor, MI.

Schweickart, Russell. "There Ain't No Graceful Way: Urination and Defecation in Zero-G." Interview with Peter Warshall. In *Space Colonies*, edited by Stewart Brand, 116–19. New York: Penguin, 1977.

Sharpe, Mitchell R. "The Human Parameter in Space Flight." NASA TM-X-57119, 1963. Available at NASA Technical Reports Server, http://ntrs.nasa.gov.

Sisakyan, N. M., ed. *Problems of Space Biology*, vol. 4. Moscow: USSR Academy of Science, 1965. NASA Technical Translation TT F-368. Available at NASA Technical Reports Server, http://ntrs.nasa.gov.

Stever, H. Guyford. Papers. Gerald R. Ford Library. Ann Arbor, MI.

Tibbitts, T. W., W. Cao, and R. M. Wheeler. *Growth of Potatoes for CELSS*. NASA Contractor Report 177646, 1994. Available at NASA Technical Reports Server, http://ntrs.nasa.gov.

Went, Frits. Papers. Record Group 3/2/6/1. Archives. Missouri Botanical Garden.

Wieland, Paul O. *Designing for Human Presence in Space: An Introduction to Environmental Control and Life Support Systems*. NASA Publication 1324. Washington, D.C.: NASA, 1994. Available at NASA Technical Reports Server, http://ntrs.nasa.gov.

Wilkins, Judd R. "Man, His Environment and Microbiological Problems of Long-Term Space Flight." NASA TMX 60422, 1967. Available at NASA Technical Reports Server, http://ntrs.nasa.gov.

Wheeler, R. M., C. L. Mackowiak, T. W. Dreschel, J. C. Sager, R. P. Prince, W. M. Knott, C. R. Hinkle, and R. F. Strayer. "System Development and Early Biomass Tests in NASA's Biomass Production Chamber." NASA Technical Memorandum TM 103494, 1990. Available at NASA Technical Reports Server, http://ntrs.nasa.gov.

Published Works

Acevedo, Sara. "In Memoriam Dr. Harold P. Klein (1921–2001)." *Origins of Life and Evolution of the Biosphere* 31, no. 6 (2001): 549–51.

Ackert, Lloyd. *Sergei Vinogradskii and the Cycle of Life: From the Thermodynamics of Life to Ecological Microbiology, 1850–1950*. Dordrecht: Springer, 2012.

Adas, Michael. *Dominance by Design: Technological Imperatives and America's Civilizing Mission*. Cambridge, MA: Belknap Press, 2006.

Allen, John. *Biosphere 2: The Human Experiment*. London: Penguin, 1991.

Allen, John. "A Brief History of the Institute of Ecotechnics' Series of 'International Meetings on Closed Ecological Systems and Biospherics.'" *Life Support and Biosphere Science* 4 (1997): 93–94.

Allen, John. *Me and the Biospheres*. Santa Fe: Synergetic Press, 2009.

Allen, Tom. *The Quest: A Report on Extraterrestrial Life*. Philadelphia: Chilton, 1965.

Alling, Abigail, and Mark Nelson. *Life under Glass: The Inside Story of Biosphere 2*. Oracle: Synergetic Press, 1993.

Ambrose, Stephen E., and Douglas G. Brinkley. *Rise to Globalism: American Foreign Policy since 1938*. London: Penguin, 1997.

Andrews, James T. *Red Cosmos: K.E. Tsiolkovskii, Grandfather of Soviet Rocketry*. College Station: Texas A&M University Press, 2009.

Andrews, James T., and Asif A. Siddiqi, eds. *Into the Cosmos: Space Exploration and Soviet Culture*. Pittsburgh: University of Pittsburgh Press, 2011.

Ankeny, Rachel. "Wormy Logic: Model Organisms as Case-Based Reasoning." In *Science without Laws: Model Systems, Cases, Exemplary Narratives,* edited by Angela N. H. Creager, Elizabeth Lunbeck, and M. Norton Wise, 46–58. Durham: Duke University Press, 2007.

Anker, Peder. "The Ecological Colonization of Space." *Environmental History* 10 (2005): 239–68.

Anker, Peder. *From Bauhaus to Ecohouse: A History of Ecological Design*. Baton Rouge: Louisiana State University Press, 2010.

Appel, Toby. *Shaping Biology: The National Science Foundation and American Biological Research, 1945–1975*. Baltimore: Johns Hopkins University Press, 2000.

Ard, Patricia. "Garbage in the Garden State: A Trash Museum Confronts New Jersey's Image." *Public Historian* 27 (2005): 57–66.

Armstrong, Neil, Edwin Aldrin, and Michael Collins, with Gene Farmer and Dora Jane Hamblin. *First on the Moon: A Voyage with Neil Armstrong, Edwin Aldrin, and Michael Collins*. Boston: Little, Brown, 1970.

Aronowsky, Leah V. "NASA and the Dream of Multispecies Spaceflight." *Environmental Humanities* 9, no. 2 (2017): 359–77.

Averner, Maurice, and Robert MacElroy. *On the Habitability of Mars: An Approach to Planetary Ecosynthesis*. Washington, D.C.: NASA National Technical Information Service, 1976.

Bailyn, Bernard. "The Historiography of the Losers." In *The Ordeal of Thomas Hutchison*, 397–408. Cambridge: Belknap, 1968.

Ballester, Antonio, E. S. Barghoorn, Daniel B. Botkin, James Lovelock, Ramon Margalef, Lynn Margulis, Juan Oro Lynn, Rusty Schweickart, David Smith, T. Swain, John Todd, Nancy Todd, and George M. Woodwell. "Ecological Considerations for Space Colonies." *Bulletin of the Ecological Society of America* 58, no. 1 (1977): 2–4.

Batov, V. A., V. N. Belyanin, I. I. Gitelson, B. G. Kovrov, F. Ya. Sid'ko, I. A. Terskov, and I. S. Yeroshin. "Dense Continuous Cultivation of Chlorella under Various Illumination Conditions." In *Problems of Space Biology,* vol. 4, edited by N. M. Sisakyan, 650–53. Moscow: USSR Academy of Science, 1965.

Belasco, Warren. "Algae Burgers for a Hungry World? The Rise and Fall of Chlorella Cuisine." *Technology and Culture* 38 (1997): 608–34.

Belasco, Warren. "Food and the Counterculture." In *Food and Global History,* edited by Raymond Grew, 273–92. Colorado: Westview, 1999.

Belasco, Warren. *Meals to Come: A History of the Future of Food*. Berkeley: University of California Press, 2006.

Benemann, John R. "Professor William J. Oswald: An Introduction." *Journal of Applied Phycology* 15, no. 2–3 (2003): 97–98.

Benford, Gregory, and George Zebrowski, eds. *Skylife: Space Habitats in Story and Science.* San Diego: Harcourt, 2000.

Benidickson, Jamie. *The Culture of Flushing: A Social and Legal History of Sewage.* Vancouver: University of British Columbia Press, 2007.

Berkner, L. V. *The Scientific Age: The Impact of Science on Society.* New Haven: Yale University Press, 1964.

Bernstein, Carl, and Bob Woodward. *All The President's Men.* New York: Warner, 1976.

Bewicke, Dhyana, and Beverly A. Potter. *Chlorella: The Emerald Food.* Berkeley: Ronin, 1984.

Bimm, Jordan. "Psychology of Space Exploration: Contemporary Research in Historical Perspective by Douglas A. Vakoch." *Isis* 103, no. 4 (2012): 812.

Birnbaum, Juliana, and Louis Fox, eds. *Sustainable Revolution: Permaculture in Ecovillages, Urban Farms, and Communities Worldwide.* Berkeley: North Atlantic, 2014.

Bizony, Piers. *How to Build Your Own Spaceship.* New York: Plume, 2009.

Bjurstedt, Hilding, ed. *Proceedings of the First International Symposium on Basic Environmental Problems of Man in Space Paris, 29 October–2 November 1962.* New York: Springer-Verlag, 1965.

Bocking, Stephen. *Ecologists and Environmental Politics: A History of Contemporary Ecology.* New Haven: Yale University Press, 1997.

Bowler, Peter J., and Iwan Rhys Morus. *Making Modern Science: A Historical Survey.* Chicago: University of Chicago Press, 2005.

Bowman, Norman. "The Food and Atmosphere Control Problem in Space Vessels." *Journal of the British Interplanetary Society* 12 (1953): 159–97.

Bova, Ben. *Mars.* New York: Bantam, 1992.

Bova, Ben. *Mars, Inc.: The Billionaire's Club.* New York: Baen, 2013.

Bozhko, Andrei, and Violetta Semyonovna Gorodinskaya. *A Year in a "Starship."* Moskva: Molodaia Gvardiia, 1975.

Brands, H. W. *American Dreams: The United States since 1945.* New York: Penguin, 2010.

Brand, Stewart, ed. *Space Colonies.* New York: Penguin, 1977.

Brin, David. "Ice Pilot." In *Project Solar Sail,* edited by Arthur C. Clarke, 53–66. New York: RoC, 1990.

Broecker, Wallace S., and Robert Kunzig. *Fixing Climate: What Past Climate Changes Reveal about the Current Threat—and How to Counter It.* New York: Hill and Wang, 2009.

Brooks, Courtney G., James M. Grimwood, and Loyd S. Swenson Jr. *Chariots of Apollo: A History of Manned Lunar Spacecraft.* Washington, D.C.: NASA, 1979.

Brower, Kenneth. *The Starship and the Canoe.* New York: Harper Colophon, 1983.

Brzezinski, Matthew. *Red Moon Rising: Sputnik and the Hidden Rivalries that Ignited the Space Age.* New York: Times, 2007.

Burgess, Colin, and Chris Dubbs. *Animals in Space: From Research Rockets to the Space Shuttle.* Dordrecht: Springer-Praxis, 2007.

Burlew, John S., ed. *Algal Culture: From Laboratory to Pilot Plant.* Washington, D.C.: Carnegie Institution of Washington Publication 600, 1953.

Burrows, William E. *This New Ocean: The Story of the First Space Age*. New York: Modern Library, 1999.

Bush, Vannevar. *Pieces of the Action*. New York: William Morrow, 1970.

Bushnell, David. "The Beginning of Research in Space Biology at the Air Force Missile Development Center, Holloman AFB, New Mexico, 1946–1952." *Quest: The History of Spaceflight Quarterly* 23, no. 1 (2016): 39–49.

Byrd, Deborah L. "The Long Road Starts Here." *Alcalde: The Official Publication of the Texas Exes* 56, no. 4 (1979): 26–29.

Catling, David C. *Astrobiology: A Very Short Introduction*. Oxford: Oxford University Press, 2013.

Campos, Luis. *Radium and the Secret of Life*. Chicago: University of Chicago Press, 2015.

Canby, Thomas Y. "Skylab, Outpost on the Frontier of Space." *National Geographic* (October 1974): 441–92.

Carson, Rachel. *Silent Spring*. New York: Houghton Mifflin, 1962.

Carter, Jimmy. *White House Diary*. New York: Farrar, Straus and Giroux, 2010.

Carter, W. Hodding. *Flushed: How the Plumber Saved Civilization*. New York: Atria, 2006.

Chaikin, Andrew. *A Man on the Moon: The Voyages of the Apollo Astronauts*. London: Penguin, 2007.

Chekhonadskiy, N. A. "Cybernetics and Space Biology." In *Problems of Space Biology*, vol. 4, edited by N. M. Sisakyan, 192–200. Moscow: USSR Academy of Science, 1965.

Chernov, V. N., and V. I. Yakovlev. "Research on Animal Flight in an Artificial Earth Satellite." In *Artificial Earth Satellites*, edited by L. V. Kurnosova, 102–18. London: Plenum, 1960.

Chuchkin, V. G., A. S. Ushakov, V. I. Rozhdestvensky, V. N. Golovin, K. S. Arbuzova, I. V. Tsvetkova, and A. V. Kostetsky. "Some Aspects of Utilization of Higher Plants as Nutrition Sources in Space Missions." In *Life Sciences and Space Research VIII: Proceedings of the Open Meeting of Working Group V at the Twelfth Plenary Meeting of COSPAR*, edited by W. Vishias and F. G. Favorite, 302–8. Amsterdam: North-Holland, 1970.

Clarke, Arthur C. *The Sands of Mars*. London: Sidgwick and Jackson, 1951.

Clarke, Arthur C. *The Promise of Space*. New York: Harper and Row, 1968.

Cloud, Wallace. "Artificial Gills." *Popular Mechanics* (December 1967): 69–72, 189.

Cohen, Lizabeth. *A Consumer's Republic: The Politics of Mass Consumption in Postwar America*. New York: Alfred A. Knopf, 2003.

Coleman, David C. *Big Ecology: The Emergence of Ecosystem Science*. Berkeley: University of California Press, 2010.

Committee on Science and Astronautics. *Astronautical and Aeronautical Events of 1962: Report of the National Aeronautics and Space Administration to the Committee on Science and Astronautics U.S. House of Representatives Eighty-Eighth Congress*. Washington, D.C.: Committee on Science and Astronautics, 1963.

Commoner, Barry. *The Closing Circle: Nature, Man, Technology*. New York: Random House, 1971.

Compton, W. David, and Charles D. Benson. *Living and Working in Space: A History of Skylab*. Washington, D.C.: NASA, 1983.

Connolly, Janis H. "Architecture." In *Isolation: NASA Experiments in Closed-Environment*

Living, edited by Helen W. Lane, Richard L. Sauer, and Daniel L. Feeback, 59–85. San Diego: American Astronautical Society, 2002.

Conway, Erik M. *Exploration and Engineering: The Jet Propulsion Laboratory and the Quest for Mars*. Baltimore: Johns Hopkins Press, 2015.

Cooke, G. Dennis. "Ecology of Space Travel." In *Fundamentals of Ecology*, by Eugene P. Odum, 497–512. Philadelphia: W. B. Saunders, 1971.

Cowan, Ruth Swartz. *More Work for Mother: The Ironies of Household Technology from the Open Hearth to the Microwave*. New York: Basic, 1983.

Craig, Campbell, and Sergey Radchenko. *The Atomic Bomb and the Origins of the Cold War*. New Haven: Yale University Press, 2008.

Creager, Angela N. H. *The Life of a Virus: Tobacco Mosaic Virus as an Experimental Model, 1930–1965*. Chicago: University of Chicago Press, 2002.

Creager, Angela N. H. *Life Atomic: A History of Radioisotopes in Science and Medicine*. Chicago: University of Chicago Press, 2013.

Creager, Angela N. H., Elizabeth Lunbeck, and M. Norton Wise, eds. *Science without Laws: Model Systems, Cases, Exemplary Narratives*. Durham: Duke University Press, 2007.

Crick, Francis. *What Mad Pursuit: A Personal View of Scientific Discovery*. New York: Basic Books, 1988.

Cringely, Robert X. *Accidental Empires: How the Boys of Silicon Valley Made Their Millions, Battled Foreign Competition, and Still Can't Get a Date*. New York: Harper Business, 1996.

Cronon, William, ed. *Uncommon Ground: Rethinking the Human Place in Nature*. New York: W. W. Norton, 1996.

Crylen, Jon. "Living in a World without Sun: Jacques Cousteau, *Homo aquaticus*, and the Dream of Dwelling Undersea." *Journal of Cinema and Media Studies* 58, no. 1 (2018), 1–23.

Curry, Helen Anne. *Evolution Made to Order: Plant Breeding and Technological Innovation in Twentieth-Century America*. Chicago: University of Chicago Press, 2016.

Curtis, Valerie. *Don't Look, Don't Touch, Don't Eat: The Science behind Revulsion*. Chicago: University of Chicago Press, 2013.

David, Leonard. "Political Acceptability of Mars Exploration: Post-1981 Observations." In *The Case for Mars*, edited by Christopher P. McKay, 35–48. San Diego: American Astronautical Society, 1985.

Dawson, Jonathan. *Ecovillages: New Frontiers for Sustainability*. Cambridge: UIT Cambridge, 2006.

Degroot, Gerard: *Dark Side of the Moon: The Magnificent Madness of the American Lunar Quest*. New York: New York University Press, 2006.

Del Duca, Michael G. "Nutrition-Waste Complex: A Pressing Problem in Manned Space Exploration." In *Conference on Nutrition in Space and Related Waste Problems*, 9–11. NASA SP-70. Washington, D.C.: NASA, 1964.

DeMeis, Richard. "Mir: A Second Sputnik." *Aerospace America* (July 1987): 24–26.

Dempster, W. F. "Methods for Measurement and Control of Leakage in CELSS and their Application and Performance in the Biosphere 2 Facility." *Advances in Space Research* 14, no. 11 (1994): 331–35.

Dempster, William F. "Engineering of Biosphere 2: Closure and Energy." *Life Support and Biosphere Science* 4 (1997): 109–16.

DeVorkin, David, *Science with a Vengeance: How the Military Created the US Space Sciences after World War II*. Dordrecht: Springer-Verlag, 1992.

Dick, Steven J., ed. *The Impact of Discovering Life Beyond Earth*. Cambridge: Cambridge University Press, 2015.

Dick, Steven J., and James E. Strick. *The Living Universe: NASA and the Development of Astrobiology*. New Brunswick, NJ: Rutgers University Press, 2004.

Dickey, Beth. "Seeding Space." *Space World* (April 1987): 15–19.

Dickson, Paul. *Sputnik: The Shock of the Century*. New York: Walker, 2011.

Douglas, Mary. *Purity and Danger: An Analysis of Concepts of Pollution and Taboo*. London: Routledge Classics, 2002.

Drexler, K. Eric. "The Canvas of the Night." In *Project Solar Sail*, edited by Arthur C. Clarke, 41–52. New York: RoC, 1990.

Duggins, Pat. *Trailblazing Mars: NASA's Next Giant Leap*. Gainesville: Florida University Press, 2010.

Dyson, Freeman. *Disturbing the Universe*. New York: Basic Books, 1979.

Eckart, Peter. *Spaceflight Life Support and Biospherics*. Dordrecht: Kluwer Academic, 1996.

Edgerton, David. *Shock of the Old: Technology and Global History Since 1900*. Oxford: Oxford University Press, 2006.

Egan, Michael. *Barry Commoner and the Science of Survival*. Cambridge, MA: MIT Press, 2007.

Egan, Michael. "Survival Science: Crisis Disciplines and the Shock of the Environment in the 1970s." *Centaurus* 59 (2017): 26–39.

Farber, Paul Lawrence. *Finding Order in Nature: the Naturalist Tradition from Linnaeus to E. O. Wilson*. Baltimore: Johns Hopkins University Press, 2000.

Finér, L., P. Aphalo, U. Kettunen, I. Leinonen, and H. Mannerkoski. "The Joensuu Dasotrons: A New Facility for Studying Shoot, Root, and Soil Processes." *Plant and Soil* 231 (2001): 137–49.

Folsome, Clair Edwin. *The Origin of Life*. San Francisco: W. H. Freeman, 1979.

Folsome, Clair Edwin, and Joe A. Hanson. "The Emergence of Materially-Closed-System Ecology." In *Ecosystem Theory and Application*, edited by Nicholas Polunin, 269–88. New York: John Wiley and Sons, 1986.

Foner, Eric. *The Story of American Freedom*. New York: W. W. Norton, 1998.

Ford, Daniel. *The Cult of the Atom: The Secret Papers of the Atomic Energy Commission*. New York: Simon and Schuster, 1982.

Forman, Paul. "The Primacy of Science in Modernity, of Technology in Postmodernity and of Ideology in the History of Technology." *History and Technology* 23 (2007): 1–152.

Frank, Adam, and Woodruff Sullivan. "Sustainability and the Astrobiological Perspective: Framing Human Futures in a Planetary Context." *Anthropocene* 5 (2014): 32–41.

Freidel, Robert. *Zipper: An Exploration in Novelty*. London: Norton, 1994.

French, C. Stacy. "Photosynthesis." In *This Is Life: Essays in Modern Biology*, edited by Willis H. Johnson and William C. Steere, 3–40. New York: Holt, Rinehart and Winston, 1962.

Friedel, Robert. "History, Sustainability, and Choice." In *Cycling and Recycling: Histories of Sustainable Practices*, edited by Ruth Oldenziel, and Helmuth Trischler, 219–25. New York: Berghahn, 2016.

Gaddis, John Lewis. *The United States and the Origins of the Cold War, 1941–1947*. New York: Columbia University Press, 1972.

Galbraith, John Kenneth. *The Affluent* Society. New York: Houghton Mifflin, 1958.

Gentry, Linnea, and Karen Liptak. *The Glass Ark: The Story of Biosphere 2*. New York: Viking, 1991.

George, Rose. *The Big Necessity: The Unmentionable World of Human Waste and Why It Matters*. New York: Picador, 2014.

Geppert, Alexander C. T., ed. *Imagining Outer Space: European Astroculture in the Twentieth Century*. London: Palgrave Macmillan, 2012.

Gerovitch, Slava. "'New Soviet Man inside Machine': Human Engineering, Spacecraft Design, and the Construction of Communism." *Osiris* 22 (2007): 135–57.

Ghamari-Tabrizi, Sharon. "Cognitive and Perceptual Training in the Cold War Man-Machine System." In *Uncertain Empire: American History of the Idea of the Cold War*, edited by Joel Isaac and Duncan Bell, 267–93. Oxford: Oxford University Press, 2012.

Ginzburg, Carlo. *The Cheese and the Worms*. London: Penguin Books, 1992.

Gitelson, Josef I. "Biological Life-Support Systems for Mars Mission." *Advances in Space Research* 12, no. 5 (1992): 167–92.

Gitelson, Josef I., Genrich M. Lisovsky, and Robert D. MacElroy. *Manmade Closed Ecological Systems*. London: Taylor and Francis, 2003.

Gitelson, Josef I., I. A. Terskov, B. G. Kovrov, G. M Lisovskii, Yu. N. Okladnikov, F. Ya. Sid'ko, I. N. Trubachev, M. P. Shilenko, S. S. Alekseev, I. M. Pan'kova, and L. S. Turranen. "Long-Term Experiments on Man's Stay in Biological Life-Support." *Advance in Space Research* 9, no. 8 (1989): 65–71.

Godwin, Robert, ed. *Apollo 9: The NASA Mission Reports*. Burlington, Ontario: Apogee, 1971.

Godwin, Robert, ed. *Apollo 8: The NASA Mission Reports*. Burlington, Ontario: Apogee, 2000.

Golueke, Clarence G., and William J. Oswald. *Biological Control of Enclosed Environments*. Berkeley: University of California Sanitary Engineering Research Laboratory, 1958.

Golueke, Clarence G., and William J. Oswald. "Closing an Ecological System Consisting of a Mammal, Algae, and Non-photosynthetic Microorganisms." *American Biology Teacher* 25, no. 7 (1963): 522–28.

Golueke, Clarence G., and William J. Oswald. "Role of Plants in Closed Systems." *Annual Review of Plant Physiology* 15 (1964): 387–408.

Golueke, Clarence G., and William J. Oswald. "The Algatron: A Novel Microbial Culture System." *Sun at Work* 11, no. 1 (1966): 3–9.

Gorbov F. D., and Novikov, M. A. "Experimental Psychological Investigation of Cosmonaut Teams." In *Problems of Space Biology*, vol. 4, edited by N. M. Sisakyan, 12–21. Moscow: USSR Academy of Science, 1965.

Gordin, Michael D. *Red Cloud at Dawn: Truman, Stalin, and the End of the Atomic Monopoly*. New York: Farrar, Straus, and Giroux, 2009.

Gorman, Alice. "The Sky Is Falling: How Skylab Became an Australian Icon." *Journal of Australian Studies* 35, no. 4 (2011): 529–46.

Gottlieb, Robert. *Forcing the Spring: The Transformation of the American Environmental Movement*. Washington, D.C.: Island Press, 2005.

Guha, Ramachandra. *How Much Should a Person Consume? Environmentalism in India and the United States.* Berkeley: University of California Press, 2006.

Günel, Gökçe. *Spaceship in the Desert: Energy, Climate Change, and Urban Design in Abu Dhabi.* Durham: Duke University Press, 2019.

Hagen, Joel B. *An Entangled Bank: The Origins of Ecosystem Ecology.* New Brunswick: Rutgers University Press, 1992.

Hagen, Joel B. "Teaching Ecology during the Environmental Age, 1965–1980." *Environmental History* 13 (2008): 704–23.

Halstead, Thora W., and F. Ronald Dutcher. "Plants in Space." *Annual Review of Plant Physiology* 38 (1987): 317–45.

Hamblin, Jacob Darwin. *Poison in the Well: Radioactive Waste in the Ocean at the Dawn of the Nuclear Age.* New Brunswick: Rutgers University Press, 2009.

Hanley, Susan B. *Everyday Things in Premodern Japan: The Hidden Legacy of Material Culture.* Berkeley: University of California Press, 1999.

Harford, James. *Korolev: How One Man Masterminded the Soviet Drive to Beat America to the Moon.* New York: John Wiley and Sons, 1997.

Hark, Ina Rae. *Star Trek.* New York: BFI/Palgrave Macmillan, 2008.

Harrison Albert A., and Mary M. Connors. "Psychological and Interpersonal Adaptation to Mars Missions." In *The Case for Mars*, edited by Christopher P. McKay, 643–54. San Diego: American Astronautical Society, 1985.

Hartman, Edwin P. *Adventures in Research: A History of the Ames Research Center, 1940–1965.* Washington D.C.: NASA, 1970.

Hecht, Gabrielle. *The Radiance of France: Nuclear Power and National Identity after World War II.* Cambridge, MA: MIT Press, 2009.

Henninger, Donald L. "Test Phases and Major Findings." In *Isolation: NASA Experiments in Closed-Environment Living,* edited by Helen W. Lane, Richard L. Sauer, and Daniel L. Feeback, 35–49. San Diego: American Astronautical Society, 2002.

Heppenheimer, T. A. *Colonies in Space.* Harrisburg, PA: Stackpole, 1977.

Hersch, Matthew H. *Inventing the American Astronaut.* New York: Palgrave Macmillan, 2012.

Heyerdahl, Thor. *The Ra Expeditions.* New York: Doubleday, 1971.

Hinghofer-Szalkay, Helmut G,. and David Moore. "Some Comments on Biological Aspects of Life Support Systems." In *Biological and Medical Research in Space: An Overview of Life Sciences Research in Microgravity*, 542–61. Edited by David Moore, Peter Bie, and Heinz Oser. Berlin: Springer-Verlag, 1996.

Hoagland, Alison K. *The Bathroom: A Social History of Cleanliness and the Body.* Santa Barbara: Greenwood, 2018.

Höhler, Sabine. *Spaceship Earth in the Environmental Age, 1960–1990.* New York: Routledge, 2016.

Holt, Nathalia. *Rise of the Rocket Girls: The Women Who Propelled Us, from Missiles to the Moon to Mars.* Boston: Back Bay, 2017.

Howell, David L. "Fecal Matters: Prolegomenon to a History of Shit in Japan." In *Japan at Nature's Edge: The Environmental Context of a Global Power*, edited by Ian Jared Miller, Julia Adeney Thomas, and Brett L. Walker, 137–51. Honolulu: University of Hawai'i Press, 2013.

Hughes, Sally Smith. *Genentech: The Beginnings of Biotech.* Chicago: University of Chicago Press, 2011.

Hughes, Thomas P. *Rescuing Prometheus: Four Monumental Projects that Changed the Modern World.* New York: Vintage, 2000.

Hughes, Thomas P. *Human-Built World: How to Think about Technology and Culture.* Chicago: University of Chicago Press, 2005.

Ingrassia, Paul. *Engines of Change: The History of the American Dream in Fifteen Cars.* New York: Simon and Schuster, 2012.

Inoue, E., Z. Uchijima, T. Saito, S. Isobe, and K. Uemura. "The 'Assimitron,' a Newly Devised Instrument for Measuring CO_2 Flux in the Surface Air Layer." *Journal of Agricultural Meteorology, Tokyo* 25 (1969): 165–71.

Irwin, James B., and William A. Emerson. *To Rule the Night: The Discovery Voyage of Astronaut Jim Irwin.* Philadelphia: A. J. Holman, 1973.

Isaac, Joel, and Duncan Bell, eds. *Uncertain Empire: American History and the Idea of the Cold War.* Oxford: Oxford University Press, 2012.

Jacobs, Jane. *Death and Life of Great American Cities.* New York: Random House, 1961.

Jagow, R. B., and R. S. Thomas. "Study of Life Support Systems for Space Missions Exceeding One Year in Duration." In *The Closed Life-Support System,* Ames Research Center, NASA SP-134, 75–144. Washington, D.C.: NASA, 1967.

Jervis, Robert. "Identity in the Cold War." In *The Cambridge History of the Cold War,* vol. 2, *Crises and Détente,* edited by Melvyn P. Leffler and Odd Arne Westad, 22–43. Cambridge: Cambridge University Press, 2011.

Johnson, Ann. *Hitting the Brakes: Engineering Design and the Production of Knowledge.* Durham: Duke University Press, 2009.

Josephson, Paul R. *Red Atom: Russia's Nuclear Power Program for Stalin to Today.* Pittsburgh: University of Pittsburgh Press, 2005.

Kaiser, David. "Cold War Requisitions, Scientific Manpower, and the Production of American Physicists after World War II." *Historical Studies in the Physical and Biological Sciences* 33, no. 1 (2002): 131–59.

Kaiser, David. *Drawing Theories Apart: The Dispersion of Feynman Diagrams in Postwar Physics.* Chicago: University of Chicago Press, 2005.

Kaiser, David, and Andrew Warwick, eds. *Pedagogy and the Practice of Science: Historical and Contemporary Perspectives.* Cambridge: Cambridge University Press, 2005.

Kaiser, David, and W. Patrick McCray, eds. *Groovy Science: Knowledge, Innovation, and American Counterculture.* Chicago: University of Chicago Press, 2016.

Keller, Evelyn Fox. *Refiguring Life: Metaphors of Twentieth-Century Biology.* New York: Columbia University Press, 1995.

Keller, Evelyn Fox. "Nature, Nurture, and the Human Genome." In *The Code of Codes: Scientific and Social Issues in the Human Genome Project,* edited by Daniel J. Kevles, and Leroy Hood, 281–99. Cambridge, MA: Harvard University Press, 1992.

Kelly, James Patrick. "Breakaway, Backdown." In *Skylife: Space Habitats in Story and Science,* edited by Gregory Benford and George Zebrowski, 97–105. New York: Harcourt, 2000.

Kelly, Kevin. *Out of Control: The New Biology of Machines, Social Systems, and the Economic World.* New York: Basic, 1995.

Kelly, Scott. *Endurance: A Year in Space, A Lifetime of Discovery*. New York: Alfred A. Knopf, 2017.

Kermode, Mark. *Silent Running*. London: British Film Institute, 2014.

Kevles, Bettyann Holtzmann. *Almost Heaven: The Story of Women in Space*. New York: Basic, 2003.

Kevles, Daniel J. *The Physicists: The History of a Scientific Community in Modern America*. Cambridge, MA: Harvard University Press, 1995.

Kevles, Daniel, and Gerald L. Geison. "The Experimental Life Sciences in the Twentieth Century." *Osiris* 10 (1995): 97–121.

Kibe, S., K. Suzuki, A. Ashida, K. Otsubo, and K. Nitta. "Controlled Ecological Life Support System-Related Activities in Japan." *Life Support Biosphere Science* 4, no. 3–4 (1997): 117–25.

Kidder, Tracy. *The Soul of a New Machine*. Boston: Little, Brown, 1981.

Kidger, Neville. "Salyut 6 Mission Report." *Spaceflight* 22, no. 2 (February 1980): 50–63.

Kilgore, De Witt Douglas. *Astrofuturism: Science, Race, and Visions of Utopia in Space*. Philadelphia: University of Pennsylvania Press, 2003.

Kimball Smith, Alice. *A Peril and a Hope: The Scientists' Movement in America, 1945–47*. Cambridge: MIT Press, 1971.

Kingsland, Sharon. *The Evolution of American Ecology*. Baltimore: Johns Hopkins University Press, 2005.

Kirensky, L.V., I. I. Gitelson, I. A. Terskov, B. G. Kovrov, G. M. Lisovsky, and Yu. N. Okladnikov. "Theoretical and Experimental Decisions in the Creation of an Artificial Ecosystem for Human Life Support in Space." *Life Sciences and Space Research* 9 (1971): 75–80.

Kirk, Andrew. *Counterculture Green: The Whole Earth Catalog and American Environmentalism*. Lawrence: University of Kansas Press, 2007.

Kirkby, Diane, and Sean Scalmer. "Social Movement, Internationalism and the Cold War: Perspectives on Labor History." *Labour History* 111 (2016): 1–11.

Kirsch, David. *The Electric Vehicle Company and the Burden of History*. Baltimore: Johns Hopkins University Press, 2000.

Klein, Naomi. *This Changes Everything: Capitalism vs. The Climate*. New York: Simon and Schuster, 2014.

Klerkx, Greg. *Lost in Space: The Fall of NASA and the Dream of a New Space Age*. New York: Pantheon, 2004.

Knott, W. M. "The Breadboard Project: A Functioning CELSS Plant Growth System." *Advances in Space Research* 12, no. 5 (1992): 45–52.

Kojevnikov, Alexei. "The Cultural Spaces of the Soviet Cosmos." In *Into the Cosmos: Space Exploration and Soviet Culture*, edited by James T. Andrews and Asif A. Siddiqi, 133–55. Pittsburgh: University of Pittsburgh Press, 2011.

Konecci, Eugene B. "Closed Ecological Systems." *Space Sciences Reviews* 6 (1966): 3–20.

Köster, Roman. "Waste to Assets: How Household Waste Recycling Evolved in West Germany." In *Cycling and Recycling: Histories of Sustainable Practices*, edited by Ruth Oldenziel and Helmuth Trischler, 168–82. New York: Berghahn Books, 2016.

Krauss, Robert W. "Discussion: Combined Photosynthetic Regenerative Systems." In

Conference on Nutrition in Space and Related Waste Problems, 289–97. Washington, D.C.: NASA, 1964.

Krauss, Robert W. "The Physiology and Biochemistry of Algae, with Space Reference to Continuous-Culture Techniques for *Chlorella*." In *Bioregenerative Systems*, 97–109. Washington, D.C.: NASA, 1968.

LaFeber, Walter. *America, Russia, and the Cold War, 1945–2002*. Boston: McGraw-Hill, 2004.

Landfester, Ulrike, Nina-Louisa Remuss, Kai-Uwe Schrogl, and Jean-Claude Worms, eds. *Humans in Outer Space: Interdisciplinary Perspectives*. Wien: Springer, 2011.

Lane, Helen W., Richard L. Sauer, and Daniel L. Feeback, eds. *Isolation: NASA Experiments in Closed-Environment Living*. San Diego: American Astronautical Society, 2002.

Laporte, Dominique. *History of Shit*. Cambridge MA: MIT Press Documents, 2000.

Launius, Roger. "Space Stations for the United States: An Idea Whose Time Has Come." *Acta Astronautica* 62 (2008): 539–55.

Launius, Roger D. "'Astronaut Envy?': The US Military's Quest for a Human Mission in Space," *Space and Defense* 4, no. 2 (2010): 61–78.

Lawton, J. H. "The Ecotron: A Controlled Environmental Facility for the Investigation of Population and Ecosystem Processes." *Philosophical Transactions: Biological Sciences* 341, no. 1296 (1993): 181–94.

Lawton, John H. "Are There General Laws in Ecology?" *OIKOS* 84 (1999): 177–92.

Lebedev, Valentin. *Diary of a Cosmonaut: 211 Days in Space*. Texas: Phytoresource Research, 1988.

Leffler, Melvyn P., and Odd Arne Westad, eds. *The Cambridge History of the Cold War*, vol. 1, *Origins*. Cambridge: Cambridge University Press, 2012.

Leslie, Stuart W. "Playing the Education Game to Win: The Military and Interdisciplinary Research at Stanford." *Historical Studies in the Physical and Biological Sciences* 18 (1987): 55–88.

Levashov, V. V. "New Aspects of Personal Hygiene." In *Problems of Space Biology*, vol. 4, edited by N. M. Sisakyan, 163–65. Moscow: USSR Academy of Science, 1965.

Levins, Richard, and Richard Lewonti. *The Dialectical Biologist*. Cambridge: Harvard University Press, 1985.

Lilienthal, David E. *Change, Hope, and the Bomb*. Princeton, NJ: Princeton University Press, 1963.

Litton, Craig. "Issue Coordinator's Comments." *Life Support and Biosphere Science* 4 (1997): 89–91.

Logsdon, John M. "Project Apollo: Americans to the Moon." In *Exploring the Unknown: Selected Documents in the History of the U.S. Civil Space Program*, vol. 6, *Space and Earth Science*, edited by John M. Logsdon, Stephen J. Garber, Roger D. Launius, and Ray A. Williamson, 378–439. Washington, D.C.: NASA, 2004.

Low, G. M. "Biological Payloads and Manned Space Flight." November 16, 1959. Document III-3, In *Exploring the Unknown: Selected Documents in the History of the U.S. Civil Space Program*, vol. 6, *Space and Earth Science*, edited by John M. Logsdon, Stephen J. Garber, Roger D. Launius, and Ray A. Williamson, 307–14. Washington, D.C.: NASA, 2004.

MacKenzie, Donald. *Knowing Machines: Essays on Technical Change*. Cambridge, MA: MIT Press, 1998.

Mailer, Norman. *Of a Fire on the Moon*. New York: Signet, 1970.

Mallove, Eugene F., and Gregory L. Matloff. *The Starflight Handbook: A Pioneers Guide to Interstellar Travel*. New York: Wiley Science, 1989.

Malone, Patrick. "The Skulking Way of War." In *Major Problems in the History of American Technology*, edited by Merritt Roe Smith and Gregory Clancy, 41–52. Boston: Houghton Mifflin, 1977.

Mandell, Humboldt C., Jr. "Space Station: The First Step." In *The Case for Mars*, edited by Christopher P. McKay, 157–70. San Diego: American Astronautical Society, 1985.

Marino, Bruno D. V., and H. T. Odum. "Biosphere 2: Introduction and Research Progress." *Ecological Engineering* 13 (1999): 3–14.

Mark, Hans. *The Space Station: A Personal Journey*. Durham: Duke University Press, 1987.

"The Mars 500 Isolation Experiment." In *Humans in Outer Space: Interdisciplinary Perspectives*, edited by Ulrike Landfester, Nina-Louisa Remuss, Kai-Uwe Schrogl, and Jean-Claude Worms, 306–8. Wien: Springer, 2011.

Mashinskiy, A., and G. Nechitaylo. "Birth of Space Plant Growing." Translated by NASA. *Tekhnika—Molodezhi* 4 (1983), 2–7.

Matusow, Allen J. *The Unraveling of America: A History of Liberalism in the 1960s*. New York: Harper, 1986.

May, Elaine Tyler. *Homeward Bound: American Families in the Cold War*. New York: Basic, 1999.

McCray, W. Patrick. *The Visioneers: How a Group of Elite Scientists Pursued Space Colonies, Nanotechnologies, and a Limitless Future*. Princeton: Princeton University Press, 2013.

McDougall, Walter. . . . *The Heavens and the Earth: A Political History of the Space Race*. New York: Basic, 1985.

McKay, Christopher P., ed. *The Case for Mars*. San Diego: American Astronautical Society, 1985.

McNeill, J. R. *Something New under the Sun: An Environmental History of the Twentieth Century*. New York: W. W. Norton, 2001.

Meleshko, G. I., and L. M. Krasotchenko. "Conditions of Carbon Nutrition of Chlorella in Intensive Cultures." In *Problems of Space Biology*, vol. 4, edited by N. M. Sisakyan, 643–49. Moscow: USSR Academy of Science, 1965.

Melosi, Martin V. *The Sanitary City: Environmental Services in Urban America from Colonial Times to the Present*. Pittsburgh: University of Pittsburgh Press, 2008.

Mendell, Wendell W. "A Personal History of the Human Exploration Initiative with Commentary on the Pivotal Role for Life Support Research." In *Biological Life Support Systems*, edited by Mark Nelson, and Gerald Soffen, 96–104. Oracle: Synergetic Press, 1990.

Mendell, Wendell W. "The Space Exploration Initiative: A Challenge to Advanced Life Support Technologies: Keynote Presentation." *Waste Management and Research* 9 (1991): 327–29.

Messeri, Lisa. *Placing Outer Space: An Earthly Ethnography of Other Worlds*. Durham: Duke University Press, 2016.

Michener, James A. *Space*. New York: Random House, 1982.

Milburn, Colin. "Nanotechnology in the Age of Posthuman Engineering: Science Fiction as Science." In *Nanoculture: Implications of the New Technoscience*, edited by N. Katherine Hayles, 109–29. Bristol: Intellect, 2004.

Milburn, Colin. "Nanowarriors: Soldier Nanotechnology and Comic Books." *Intertexts* 9 (2005): 77–103.

Milburn, Colin. *Nanovision: Engineering the Future*. Durham, NC: Duke University Press, 2008.

Mindell, David A. *Digital Apollo: Human and Machine in Spaceflight*. Cambridge, MA: MIT Press, 2011.

Misa, Thomas J. "The Compelling Tangle of Modernity and Technology." In *Modernity and Technology*, edited by Thomas J. Misa, Philip Brey, and Andrew Feenberg, 1–30. Cambridge, MA: MIT Press, 2003.

Mody, Cyrus C. M. "Santa Barbara Physicists in the Vietnam Era." In *Groovy Science: Knowledge, Innovation, and American Counterculture*, edited by David Kaiser and W. Patrick McCray, 70–107. Chicago: University of Chicago Press, 2016.

Mody, Cyrus C. M. "Square Scientists and the Excluded Middle." *Centaurus* 59 (2017): 58–71.

Morange, Michel. *Life Explained*. New Haven: Yale University Press, 2008.

Morgan, Karl Z. "Radiation Protection." In *Nuclear Reactors for Industry and Universities*, edited by Ernest Henry Wakefield, and Clifford K. Beck, 47. Pittsburgh: Instruments, 1954.

Muenger, Elizabeth A. *Searching the Horizon: A History of Ames Research Center, 1940–76*. Washington, D.C.: NASA Scientific and Technical Informational Brach, 1985.

Muir-Harmony, Teasel. *Apollo to the Moon: A History in 50 Objects*. Washington, D.C.: National Geographic, 2018.

Mukherjee, Siddhartha. *The Gene: An Intimate History*. New York: Scribner, 2016.

Mullane, Mike. *Do Your Ears Pop in Space? And 500 Other Surprising Questions about Space Travel*. New York: Wiley, 1997.

Mullane, Mike. *Riding Rockets: The Outrageous Tales of a Space Shuttle Astronaut*. New York: Scribner, 2006.

Mulligan, Martin, and Stuart Hill. *Ecological Pioneers: A Social History of Australian Ecological Thought and Action*. Cambridge: Cambridge University Press, 2001.

Munns, David P. D. *A Single Sky: How an International Community Forged the Science of Radio Astronomy*. Cambridge, MA: MIT Press, 2013.

Munns, David P. D. "'The Awe in which Biologists Hold Physicists': Frits Went's First Phytotron at Caltech, and an Experimental Definition of the Biological Environment." *History and Philosophy of the Life Sciences* 36, no. 2 (2014): 209–31.

Munns, David P. D. *Engineering the Environment: Phytotrons and the Quest to Control Climate in the Cold War*. Pittsburgh: University of Pittsburgh Press, 2017.

Munns, David P. D., and Kärin Nickelsen. "To Live among the Stars: Artificial Environments in the Early Space Age." *History and Technology* 33 (2018): 272–99.

Murphy, Guy. *Mars: A Survival Guide*. Australia: ABC Books, 2010.

Murphy, Michelle. *Sick Building Syndrome and the Problem of Uncertainty: Technoscience, Environmental Politics and Women Workers*. Durham: Duke University Press, 2006.

Müller-Wille, Staffan, and Hans-Jörg Rheinberger. *A Cultural History of Heredity*. Chicago: University of Chicago Press, 2012.

Müntz, Klaus. "Die Massenkultur von Kleinalgen. Bisherige Ergebnisse und Probleme." *Die Kulturpflanze* 15 (1967): 311–50.

Myers, Jack E. "Basic Remarks on the Use of Plants as Biological Gas Exchangers in a Closed System." *Journal of Aviation Medicine* 25, no. 4 (1954): 407–11.

Myers, Jack E. "Combined Photosynthetic Regenerative Systems." In *Conference on Nutrition in Space and Related Waste Problems*, 283–87. Washington, D.C.: NASA, 1964.

Nader, Ralph. *Unsafe at Any Speed: The Designed-In Dangers of the American Automobile.* New York: Grossman, 1965.

Neal, Valerie. *Spaceflight in the Shuttle Era and Beyond: Redefining Humanity's Purpose in Space.* New Haven: Yale University Press, 2017.

Nelkin, Dorothy. *The University and Military Research: Moral Politics at M.I.T.* Ithaca: Cornell University Press, 1972.

Nelson, Amy. "Cold War Celebrity and the Courageous Canine Scout." In *Into the Cosmos: Space Exploration and Soviet Culture*, edited by James T. Andrews and Asif A. Siddiqi, 133–55. Pittsburgh: University of Pittsburgh Press, 2011.

Nelson, Mark. "Biotechnology of Space Biospheres." In *Fundamentals of Space Biology*, edited by Makoto Asashima, and George M. Malacinski, 185–200. Tokyo: Japan Scientific Society Press, 1990.

Nelson, Mark. *The Wastewater Gardener: Preserving the Planet One Flush at a Time.* Santa Fe: Synergetic Press, 2014.

Nelson, Mark. *Pushing Our Limits: Insights from Biosphere 2.* Tucson: University of Arizona Press, 2018.

Nelson, Mark, and Gerald Soffen, eds. *Biological Life Support Systems.* Oracle: Synergetic Press, 1990.

Nelson, Mark, and Gerald Soffen. "Foreword." In *Biological Life Support Systems*, edited by Mark Nelson and Gerald Soffen, vii–ix. Oracle: Synergetic Press, 1990.

Nelson, Mark, John P. Allen, and William Dempster. "Biosphere 2: A Prototype Project for a Permanent and Evolving Life Support System for Mars Base." *Advances in Space Research* 12, no. 5 (1992): 211–17.

Nelson, Mark, William Dempster, and John P. Allen. "The Water Cycle in Closed Ecological Systems: Perspectives from the Biosphere 2 and Laboratory Biosphere System." *Advances in Space Research* 44 (2009): 1404–12.

Nickelsen, Kärin. "The Construction of a Scientific Model: Otto Warburg and the Building-Block-Strategy." *Studies in History and Philosophy of Biological and Biomedical Sciences* 40 (2009): 73–86.

Nickelsen, Kärin. *Explaining Photosynthesis: Models of Biochemical Mechanisms, 1840–1960.* Dordrecht: Springer, 2015.

Nickelsen, Kärin. "The Organism Strikes Back: *Chlorella* Algae and Their Impact on Photosynthesis Research, 1920s–1960s." *History and Philosophy of the Life Sciences* 39, no. 9 (2017): 1–22.

Nickelsen, Kärin. "Physicochemical Biology and Knowledge Transfer: The Study of the Mechanism of Photosynthesis between the Two World Wars." *Journal of the History of Biology*, Special Issue (2019): 1–29. DOI:10.1007/s10739-019-9559-x.

Nickelsen, Kärin, and Govindjee. *The Maximum Quantum Yield Controversy: Otto Warburg and the "Midwest Gang."* Bern: Bern Studies in the History and Philosophy of Science, 2011.

Oberg, James E. "Russians to Mars?" In *The Case for Mars*, edited by Christopher P. McKay, 73–78. San Diego: American Astronautical Society, 1985.

Oberg, James E., and Alcestis R. Oberg. *Pioneering Space: Living on the Next Frontier*. New York: McGraw-Hill, 1986.

Odum, Eugene P. *Ecology: A Bridge between Science and Society*. Sunderland, MA: Sinauer Associates, 1997.

Oldenziel, Ruth, and Karin Zachmann, eds. *Cold War Kitchen: Americanization, Technology, and European Users*. Cambridge, MA: MIT Press, 2009.

Oliver, Donna C. "Psychological Effects of Isolation and Confinement of a Winter-Over Group at McMurdo Station, Antarctica." In *From Antarctica to Outer Space: Life in Isolation and Confinement*, edited by A. A. Harrison, Y. A. Clearwater, and C. P. McKay, 217–27. New York: Springer-Verlag, 1991.

Olson, Valerie. *Into the Extreme: U.S. Environmental Systems and Politics beyond Earth*. Minneapolis: University of Minnesota Press, 2018.

O'Neill, Gerard K. *2081: A Hopeful View of the Human Future*. New York: Simon and Schuster, 1981.

Oreskes, Naomi, and John Krige, eds. *Science and Technology in the Global Cold War*. Cambridge MA: MIT Press, 2014.

Osborn, Henry Fairfield, Jr. *Our Plundered Planet*. Boston: Little, Brown, 1948.

Osborn, Henry Fairfield, Jr. *The Limits of the Earth*. Boston: Little, Brown, 1953.

Oswald, William J. "The Coming Industry of Controlled Photosynthesis." *American Journal of Public Health* 52 (1962): 235–42.

Oswald, W. J., and Clarence G. Golueke. "Man in Space—He Takes along His Waste Problem!" *Wastes Engineering* (September 1961): 456–59.

Oswald, W. J., and Clarence G. Golueke. "Environmental Control Studies with a Closed Ecological System." *Proceedings of the 8th Annual Meeting of the Institute of Environmental Sciences* 8 (1962): 183–91.

Oswald, W. J., Clarence G. Golueke, and Donald O. Horning. "Closed Ecological Systems." *Journal of the Sanitary Engineering Division, ASCE* SA4 (August 1965): 23–46.

Ott, Charles. "Waste Management for Closed Environments." In *Conference on Nutrition in Space and Related Waste Problems*, 97–102. Washington, D.C.: NASA, 1964.

Packham, Nigel J. "The Lunar-Mars Life Support Test Project: the Crew Perspective." In *Isolation: NASA Experiments in Closed-Environment Living*, edited by Helen W. Lane, Richard L. Sauer, and Daniel L. Feeback, 17–34. San Diego: American Astronautical Society, 2002.

Paglen, Trevor. *Blank Spots on the Map: The Dark Geography of the Pentagon's Secret World*. New York: Dutton, 2009.

Paine, Thomas O. "A Timeline for Martian Pioneers." In *The Case for Mars*, edited by Christopher P. McKay, 3–21. San Diego: American Astronautical Society, 1985.

Paine, Thomas O. "Biospheres and Solar System Exploration." In *Biological Life Support Systems*, edited by Mark Nelson, and Gerald Soffen, 1–11. Oracle: Synergetic Press, 1990.

Panel for the Plant Sciences. *The Plant Sciences Now and In the Coming Decade: A Report on the Status, Trends, and Requirements of Plant Sciences in the United States*. Washington, D.C.: National Academy of Sciences, 1966.

Parker, Martin. "Managing Space, Organising the Sublime." In *Humans in Outer Space: Interdisciplinary Perspectives,* edited by Ulrike Landfester, Nina-Louisa Remuss, Kai-Uwe Schrogl, and Jean-Claude Worms, 28–38. Wien: Springer, 2011.

Peake, Thomas R. *Keeping the Dream Alive: A History of the Southern Christian Leadership Conference from King to the Nineteen-Eighties*. New York: Peter Lang, 1987.

Peake, Tim. *Ask An Astronaut.* London: Century, 2017.

Perrin, Noel. *Giving Up the Gun: Japan's Reversion to the Sword, 1653–1879*. Boston: David Godine, 1988.

Phelps, J. Alfred. *They Had a Dream: The Story of African-American Astronauts*. Navato, CA: Presidio, 1994.

Pickering, Andrew. *The Cybernetic Brain: Sketches of Another Future*. Chicago: University of Chicago Press, 2010.

Pittendrigh, Colin S., Wolf Vishniac, and J. P. T. Pearman, eds. *Biology and the Exploration of Mars*. Washington, D.C.: National Academy of Sciences, 1966.

Pitts, Jon A. *The Human Factor: Biomedicine in the Manned Space Program to 1980*. Washington D.C.: NASA, 1985.

Plaxco, Kevin W., and Michael Gross. *Astrobiology: A brief Introduction*. Baltimore: Johns Hopkins University Press, 2006.

Pogue, William R. *How Do You Go to the Bathroom in Space?* New York: Tor, 1999.

Poiger, Uta G. *Jazz, Rock, and Rebels: Cold War Politics and American Culture in a Divided Germany*. Berkeley: University of California Press, 2000.

Pokrovsky, A. A., and A. S. Ushakov. "Problems of Space Nutrition of Man." In *Life Sciences and Space Research VIII: Proceedings of the Open Meeting of Working Group V at the Twelfth Plenary Meeting of COSPAR*, edited by W. Vishias and F. G. Favorite, 261–64. Amsterdam: North-Holland, 1970.

Poole, Robert. *Earthrise: How Man First Saw the Earth*. New Haven: Yale University Press, 2008.

Portree, David S. F. *Humans to Mars: Fifty Years of Mission Planning, 1950–2000*. Washington, D.C.: NASA, 2001.

Poynter, Jane. *The Human Experiment: Two Years and Twenty Minutes inside Biosphere 2*. New York: Thunder's Mouth Press, 2006.

Poynter, Jane, and D. Bearden. "Biosphere 2: A Closed Bioregenerative Life Support System, an Analog for Long Duration Space Missions." In *Plant Production in Closed Ecosystems: The International Symposium on Plant Production in Closed Ecosystems Held in Narita, Japan, August 26–29, 1996,* edited by E. Goto, K. Kurata, M. Hayashi, and S. Sase, 263–77. Dordrecht: Kluwer, 1997.

"Quails in Space." *USSR Space Life Sciences Digest* 13 (1987): 47.

Rasmussen, Nicolas. *Gene Jockeys: Life Science and the Rise of Biotech Enterprise*. Baltimore: Johns Hopkins University Press, 2014.

Rathje, William, and Cullen Murphy. *Rubbish! The Archeology of Garbage*. New York: HarperCollins, 1992.

Reid, Donald. *Paris Sewers and Sewermen: Realities and Representations*. Cambridge, MA: Harvard University Press, 1993.

Reider, Rebecca. *Dreaming the Biosphere: The Theater of All Possibilities*. New Mexico University Press, 2009.

Reynolds, Terry S., and Theodore Bernstein. "Edison and 'The Chair.'" *IEEE Technology and Society Magazine* (March 1989): 19–28.

Richers, Julia. "Remembering the Soviet Space Program." *Kritika* 18 (2017): 843–47.

Ridley, Matt. *Genome: The Autobiography of a Species in 23 Chapters*. New York: Perennial, 2000.

Roach, Mary. *Packing for Mars: The Curious Science of Life in the Void*. New York: W. W. Norton, 2010.

Robinson, Kim Stanley. *Red Mars*. New York: Bantam, 1993.

Robinson, Kim Stanley. *Aurora*. London: Orbit, 2015.

Roll-Hansen, Nils. *The Lysenko Effect: The Politics of Science*. New York: Humanity, 2005.

Rome, Adam. "Give Earth a Chance: The Environmental Movement and the Sixties." *Journal of American History* 90, no. 2 (2003): 525–54.

Rome, Adam. *The Genius of Earth Day: How a 1970 Teach-In Unexpectedly Made the First Green Generation*. New York: Hill and Wang, 2013.

Rozwadowski, Helen M. *Vast Expanses: A History of the Oceans*. London: Reaktion, 2018.

Rudolph, John. *Scientists in the Classroom: The Cold War Reconstruction of American Science Education*. New York: Palgrave, 2002.

Sabanas, Mitchell. *Closed Ecological Life-Support Unit for Primates*. Berkeley: Lawrence Radiation Laboratory, 1962.

Sagan, Carl. *Pale Blue Dot: A Vision of the Human Future in Space*. New York: Ballantine, 1997.

Sagan, Dorion. *Biospheres*. New York: Bantam, 1990.

Sagan, Dorion, and Lynn Margulis. *Biospheres: From Earth to Space*. Berkeley: Enslow, 1989.

Salisbury, Frank B. "Controlled Environment Life Support Systems (CELSS): A Prerequisite for Long-Term Space Studies." In *Fundamentals of Space Biology*, edited by Makoto Asashima and George M. Malacinski, 171–83. Tokyo: Japan Scientific Society Press, 1990.

Salisbury, Frank B., Josef I. Gitelson, and Genry M. Lisovsky. "BIOS-3: Siberian Experiments in Bioregenerative Life Support." *BioScience* 47, no. 9 (1997): 575–85.

Schrödinger, Erwin. *What Is Life?* Cambridge: Cambridge University Press, 2012.

Schmid, Sonja D. *Producing Power: The Pre-Chernobyl History of the Soviet Nuclear Industry*. Cambridge, MA: MIT Press, 2015.

Schmitt, Harrison H. "A Millenium Project: Mars 2000." In *The Case for Mars*, edited by Christopher P. McKay, 23–31. San Diego: American Astronautical Society, 1985.

Science Policy Research Division. *Review of the Soviet Space Program*. Washington, D.C.: Report of the Committee on Science and Astronautics, 1967.

Scientific American. *The Biosphere*. San Francisco: W. H. Freeman, 1970.

Scranton, Roy. *Learning to Die in the Anthropocene*. San Francisco: City Lights, 2015.

Seedhouse, Erik. *Martian Outpost: The Challenge of Establishing a Human Settlement on Mars*. Berlin: Springer, 2009.

Sellers, Christopher C. *Hazards of the Job: From Industrial Waste to Environmental Health Science*. Chapel Hill: University of North Carolina Press, 1997.

Schlosser, Eric. *Command and Control: Nuclear Weapons, the Damascus Incident, and the Illusion of Safety*. New York: Penguin, 2014.

Sharpe, Mitchell R. *Living in Space: The Astronaut and His Environment*. London: Doubleday, 1969.

Shelef, G., M. Sabanas, and W. J. Oswald. "An Improved Algatron Reactor for Photosynthetic Life Support Systems." *Proceeds of the 14th Annual Meeting of the Institute of Environmental Sciences* 14 (1968): 1–8.

Shelton, William. *Soviet Space Exploration: The First Decade*. New York: Washington Square Press, 1968.

Shepelev, Yevgeny. "Some Aspects of Human Ecology in Closed Systems with Recirculation of Substances." In *Problems of Space Biology*, vol. 4, edited by N. M. Sisakyan, 166–75. Moscow: USSR Academy of Science, 1965.

Shetterly, Margot Lee. *Hidden Figures: The American Dream and the Untold Story of the Black Women Mathematicians Who Helped Win the Space Race*. New York: William Morrow, 2016.

Shiva, Vandana: *Who Really Feeds the World? The Failure and Agribusiness and the Promise of Agroecology*. Berkeley: North Atlantic, 2016.

Siddiqi, Asif A. *Challenge to Apollo: The Soviet Union and the Space Race, 1945–1974*. Washington, D.C.: NASA, 2000.

Siddiqi, Asif A. *Sputnik and the Soviet Space Challenge*. Gainesville: University Press of Florida, 2003.

Siddiqi, Asif A. *The Red Rockets' Glare: Spaceflight and the Soviet Imagination, 1857–1957*. Cambridge: Cambridge University Press, 2010.

Siddiqi, Asif A. "Imagine the Cosmos: Utopians, Mystics, and the Popular Culture of Spaceflight in Revolutionary Russia." In *Russian Science Fiction Literature and Cinema: A Critical Reader*, edited by Anindita Banerjee, 79–116. Boston: Academic Studies, 2018.

Silent Running. Directed by Douglas Trumbull. Written by Deric Washburn, Michael Cimino, and Steven Bochco. Universal Pictures, 1972.

Siniawer, Eiko Maruko. *Waste: Consuming Postwar Japan*. Ithaca: Cornell University Press, 2018.

Simons, Dana. "Waste Not, Want Not: Excrement and Economy in Nineteenth Century France." *Representations* 96 (2006): 73–98.

Sissakian, N. M. "Contribution of the U.S.S.R. to the Exploration of Outer Space." In *Proceedings of the First International Symposium on Basic Environmental Problems of Man in Space Paris, 29 October–2 November 1962*, edited by Hilding Bjurstedt, 22–34. New York: Springer-Verlag, 1965.

Slavin, Thomas J., and Melvin W. Oleson. "Technology Tradeoffs Related to Advanced Mission Waste Processing." *Waste Management and Research* 9 (1991): 401–14.

Smith, Crosbie, and M. Norton Wise. *Energy and Empire: A Biographical Study of Lord Kelvin*. Cambridge: Cambridge University Press, 2009.

Smith, Gina. *The Genomics Age*. New York: American Management Association, 2005.

Smith, Jean Edward. *Eisenhower in War and Peace*. New York: Random House, 2013.

Smith, Robert W. *The Space Telescope: A Study of NASA, Science, Technology, and Politics*. Cambridge: Cambridge University Press, 1993.

"Space Colonization and Energy Supply to the Earth: Testimony of Dr. Gerard K. O'Neill before the Sub-committee on Space Science and Applications of the Committee on Science and Technology, United States House of Representatives, July 23, 1975." In *Space Colonies*, edited by Stewart Brand, 12–21. New York: Penguin, 1977.

Space 1999. Created by Gerry Anderson and Sylvia Anderson. Group 3 (ITV), 1975–77.

Spigel, Lynn. "White Flight." In *The Revolution Wasn't Televised: Sixties Television and Social Conflict*, edited by Lynn Spigel and Michael Curtin, 47–71. London: Routledge, 1997.

Spigel, Lynn. *Welcome to the Dreamhouse: Popular Media and Postwar Suburbs*. Durham: Duke University Press, 2001.

Stadler, Max. "Models, the Cell, and the Reformations of Biological Science, 1920–1960." PhD dissertation, Imperial College London, 2010.

Stanković, Bratislav. "2001: A Plant Space Odyssey." *Trends in Plant Science* 6, no. 12 (2001): 591–93.

Steinbuch, K. "Man or Automaton in Space." In *Proceedings of the First International Symposium on Basic Environmental Problems of Man in Space Paris, 29 October–2 November 1962*, edited by Hilding Bjurstedt, 473–92. New York: Springer-Verlag, 1965.

Stephanson, Anders. "Cold War Degree Zero." In *Uncertain Empire: American History and the Idea of the Cold War*, 19–49. Edited by Joel Isaac and Duncan Bell. Oxford: Oxford University Press, 2012.

Stradling, David. *Smokestacks and Progressives: Environmentalists, Engineers, and Air Quality in America, 1881–1951*. Baltimore: Johns Hopkins University Press, 1999.

Strasser, Susan. *Waste and Want: A Social History of Trash*. New York: Metropolitan, 1999.

Sugrue, Thomas. *The Origins of the Urban Crisis: Race and Inequality in Postwar Detroit*. Princeton: Princeton University Press, 2005.

Sullivan, W. T., III, and J. A. Baross, eds. *Planets and Life: The Emerging Science of Astrobiology*. Cambridge: Cambridge University Press, 2007.

Sylvester, Roshanna P. "She Orbits over the Sex Barrier: Soviet Girls and the Tereshkova Moment." In *Into the Cosmos: Space Exploration and Soviet Culture*, edited by James T. Andrews and Asif A. Siddiqi, 133–55. Pittsburgh: University of Pittsburgh Press, 2011.

Tarr, Joel A. "From City to Farm: Urban Wastes and the American Farmer." *Agricultural History* 49 (1975): 590–612.

Tarr, Joel A., James McCurley III, Francis C. McMichael, and Terry Yosie. "Water and Wastes: A Retrospective Assessment of Wastewater Technology in the United States." *Technology and Culture* 25 (1984), 226–63.

Taub, Frieda B. "Closed Ecological Systems." *Annual Review of Ecology and Systematics* 5 (1974): 139–60.

Teller, Edward. "Water Generation in Space." In *Conference on Nutrition in Space and Related Waste Problems*, 175–78. Washington, D.C.: NASA, 1964.

Thompson, E. P. *The Making of the English Working Class*. New York: Penguin, 1978.

Thompson, Michael. *Rubbish Theory: The Creation and Destruction of Value*. London: Pluto Press, 2017.

Tibbe, Matthew. *No Requiem for the Space Age: The Apollo Moon Landings and American Culture*. Oxford: Oxford University Press, 2014.

Tibbitts, T. W., and D. L. Henninger. "Food Production in Space: Challenges and Perspectives." In *Plant Production in Closed Ecosystems: The International Symposium on Plant Production in Closed Ecosystems Held in Narita, Japan, August 26–29, 1996,* edited by E. Goto, K. Kurata, M. Hayashi, and S. Sase, 189–203. Dordrecht: Kluwer, 1997.

Timmins, Mark. "From Space Suits to Space Couture: A New Aesthetic." In *Humans in Outer Space: Interdisciplinary Perspectives,* edited by Ulrike Landfester, Nina-Louisa Remuss, Kai-Uwe Schrogl, Jean-Claude Worms, 183–203. Wien: Springer, 2011.

Tone, Andrea. "Making Room for Rubbers: Gender, Technology and Birth Control before the Pill." *History and Technology* 18 (2002): 51–76.

Trim, Henry. "A Quest for Permanence: The Ecological Visioneering of John Todd and the New Alchemy Institute." In *Groovy Science: Knowledge, Innovation, and American Counterculture,* edited by David Kaiser and W. Patrick McCray, 142–71. Chicago: University of Chicago Press, 2016.

Tsvetkova, Ni. V., Yu. I. Shaydarov, and V. M. Abramova. "Special Features of Plant Feeding under Conditions of Aeroponic Cultivation for a Closed System." In *Problems of Space Biology,* vol. 4, edited by N. M. Sisakyan, 637–642. Moscow: USSR Academy of Science, 1965.

Turkina, Olesya. *Soviet Space Dogs.* London: FUEL, 2014.

Turner, Fred. *From Counterculture to Cyberculture: Steward Brand, the Whole Earth Network, and the Rise of Digital Utopianism.* Chicago: University of Chicago Press, 2006.

Van der Ryn, Sim. *The Toilet Papers: Recycling Waste and Conserving Water.* Santa Barbara: Capra Press, 1978.

Vaulina, E. N., and Anikeeva, I. D. "The Influence of Spaceflight on *Chlorella.*" In *Life Sciences and Space Research VIII: Proceedings of the Open Meeting of Working Group V at the Twelfth Plenary Meeting of COSPAR,* edited by W. Vishias and F. G. Favorite, 12–18. Amsterdam: North-Holland, 1970.

Velminski, Wladimir. *Homo Sovieticus: Brain Waves, Mind Control, and Telepathic Destiny.* Translated by Erik Butler. Cambridge, MA: MIT Press, 2017.

Vidal, Gore. "The State of the Union." In *Matters of Fact and Fiction: Essays, 1973–1976,* 265–85. New York: Heinemann, 1977.

Volti, Rudi. "Why Internal Combustion?" *American Heritage of Invention and Technology* (Fall 1990): 42–47.

von Puttkamer, Jesco. "Beyond the Space Station." In *The Case for Mars,* edited by Christopher P. McKay, 171–206. San Diego: American Astronautical Society, 1985.

Voskoboynikov, Fred. *The Psychology of Effective Management: Strategies for Relationship Building.* New York: Routledge, 2017.

Wagnleitner, Reinhold. *Coca-Colonization and the Cold War: The Cultural Mission of the United States in Austria after the Second World War.* Chapel Hill: University of North Carolina Press, 1994.

Waisel, Yoav, Amram Eschel, and Uzi Kafkafi, eds. *Plant Roots: The Hidden Half.* New York: Marcel Decker, 2002.

Walker, Jeremy, and Granjou, Céline. "MELiSSA the Minimal Biosphere: Human Life, Waste, and Refuge in Deep Space." *Futures* 92 (2017): 59–69.

Ware, Randolph H., and Chandler, Philip P. "The Civilian Space Program: A Washington Perspective." In *The Case for Mars,* edited by Christopher P. McKay, 49–64. San Diego: American Astronautical Society, 1985.

Watson, James D. "A Personal View of the Project." In *The Code of Codes: Scientific and Social Issues in the Human Genome Project,* edited by Daniel J. Kevles and Leroy Hood, 164–73. Cambridge, MA: Harvard University Press, 1992.

Watson, James D. *Girls, Genes, and Gamow: After the Double Helix*. New York: Knopf, 2003.

Weir, Andy. *Artemis*. New York: Crown, 2007.

Weir, Andy. *The Martian*. New York: Broadway, 2014.

Weitekamp, Margaret A. *Right Stuff, Wrong Sex: America's First Women in Space Program*. Baltimore: Johns Hopkins University Press, 2004.

Wellerstein, Alex. "Remembering Laika: Space Dog and Soviet Hero." *The New Yorker,* November 3, 2017.

Went, Frits. "The Earhart Plant Research Laboratory." *Engineering and Science* (March 1952): 14–18.

Westad, Odd Arne. *The Global Cold War: Third World Interventions and the Making of our Times*. Cambridge: Cambridge University Press, 2007.

Westwick, Peter. *Into the Black: JPL and the American Space Program, 1976–2004*. New Haven: Yale University Press, 2007.

Westwick, Peter, ed. *Blue-Sky Metropolis: The Aerospace Industry in Southern California*. Berkeley: Huntington Library and University of California Press, 2012.

Wharton, Robert A., David T. Smeroff, and Maurice M. Averner. "Algae in Space." In *Algae and Human Affairs,* edited by Carole A. Lembi and J. Robert Waaland, 485–509. Cambridge: Cambridge University Press, 1988.

Wheeler, Raymond M. "Bioregenerative Life Support and Nutritional Implications for Planetary Exploration." In *Nutrition in Spaceflight and Weightlessness Models,* edited by Helen W. Lane and Dale A. Schoeller, 41–67. Boca Raton: CRC, 2000.

Wilkinson, Ronald S., and John F. Buydos. *Aeronautical and Astronautical Resources of the Library of Congress*. Washington, D.C.: Library of Congress, 2007.

Wilson, Edward O. *Naturalist*. Washington, D.C.: Island Press, 1994.

Wisnioski, Matthew H. *Engineers for Change: Competing Vision of Technology in 1960s America*. Cambridge, MA: MIT Press, 2012.

Woese, Carl. "A New Biology for a New Century." *Microbiology and Molecular Biology Reviews* 68, no. 2 (2004): 173–86.

Wolfe, Audra J. *Competing with the Soviets: Science, Technology, and the States in Cold War America*. Baltimore: Johns Hopkins Press, 2013.

Wolfe, Audra J. *Freedom's Laboratory: The Cold War Struggle for the Soul of Science*. Baltimore: Johns Hopkins University Press, 2018.

Wolfe, Tom. *The Right Stuff.* New York: Farrar, Straus, and Giroux, 1979.

Worster, Donald. *The Good Muck: Toward an Excremental History of China*. Munich: RCC Perspectives, 2017.

Wynn, Graeme. "Foreword." In *The Culture of Flushing: A Social and Legal History of Sewage,* by Jamie Benidickson, v–viii. Vancouver: University of British Columbia Press, 2007.

Yensen, Nicholas P., and Karl Y. Biel. "Criticality Concept and Some Principles for Sustainability in Closed Biological Systems and Biospheres." In *Complex Biological Systems: Adaptation and Tolerance to Extreme Environments*, edited by Irina R. Fomina, Karl Y. Biel, and Vladislav G. Soukhovolsky, 29–71. Hoboken: John Wiley and Sons, 2018.

Yi, Doogab. *The Recombinant University: Genetic Engineering and the Emergence of Stanford Biotechnology*. Chicago: University of Chicago Press, 2015.

Zabel, P., M. Bamsey, D. Schubert, and M. Tajmar. "Review and Analysis of Over 40 Years of Space Plant Growth Systems." *Life Sciences in Space Research* 10 (2016): 1–16.

Zallen, Doris. "The 'Light' Organism for the Job: Green Algae and Photosynthesis Research." *Journal for the History of Biology* 26 (1993): 269–79.

Zic-Fuchs, Milena, and Jean-Pierre Swings. "Preface." In *Humans in Outer Space: Interdisciplinary Perspectives,* edited by Ulrike Landfester, Nina-Louisa Remuss, Kai-Uwe Schrogl, and Jean-Claude Worms, v–vii. Wien: Springer, 2011.

Zubrin, Robert. "The Significance of the Martian Frontier." 1994. Accessed November 25, 2016. http://www.nss.org/.

Zubrin, Robert. *Mars on Earth: The Adventures of Space Pioneers in the High Artic*. New York: Jeremy P. Tarcher/Penguin, 2003.

INDEX